Linear Algebra: Theorems and Applications

Claire Montemar

ARCLER
PRESS

www.arclerpress.com

Linear Algebra: Theorems and Applications

Claire Montemar

Arcler Press
2010 Winston Park Drive,
2nd Floor
Oakville, ON L6H 5R7
Canada
www.arclerpress.com
Tel: 001-289-291-7705

001-905-616-2116

Fax: 001-289-291-7601

Email: orders@arclereducation.com

ISBN: 978-1-77361-212-6 (Hardcover)

About the Author

Claire Montemar is a licensed Chemical Engineer with a real passion for teaching, learning and taking care of the environment. She obtained her degree from the University of the Philippines in 1999. She has worked in the manufacturing industry for a few years but spent most of her career as an academic tutor/instructor for elementary, high school and college students. Her forte is in the areas of mathematics and science, particularly algebra and chemistry. She currently works as a full-time freelancer, predominantly serving the needs of clients in the education industry.

Contents

Preface

Welcome to the beginning of this book -- an algebra book that shows and explains further topics than those that are covered in a primary course on linear algebra. Such course usually leaves the student with the question "Why in the world do I have to go through linear algebra?" which is absolutely understandable. Algebra, and other sciences in general, are not entities that one just needs to pass as a course. Sciences are masters that stay with you and are always there whenever you need them, that in some occasions we do not understand that those masters are something else.

Before talking about the importance of algebra, let us first clear some details about this text. This book does not only contain a set of complicated equations with heavy definitions about abstract entities and problems that arise in several fields of mathematics, which in many other texts, makes you feel like you're reading Mandarin. This book also contains clear and deductive explanations from different perspectives about those set of complicated equations, with the hope of making them accessible and easy to assimilate for the reader.

Algebra is the science of representing life through equations while keeping it simple. The reason for writing this book was, at first, the decision to spread knowledge of such a great science that is usually feared by many young students and ignored by others, when in reality Algebra is a body of knowledge so beautiful within itself, as well as necessary dealing with or approaching many mathematical situations. As this book is being developed, another reason to continue to write it arose – there is a lack of easy-to-understand texts, which are supposed to be the high end books that today's generation are obligated to read, in order to understand the subjects they should know to continue their formation in their respective careers.

This book is dedicated to those who really want to learn linear algebra. But be warned, the journey is not promised to be easy as it will require dedication, honesty and time from the reader; but the journey will also be nurturing and revealing in the sense that at the end of each chapter, the reader will be able to say "I've got it!" in much the same way when Archimedes said "Eureka!" which means "I've found it!" In addition, this book also contains techniques and algorithms, along with their derivations and examples, which are used in the real world, thus helping the reader become a well prepared professional.

Many texts, articles and lessons were used as resources for this book, however, the most used resource in the development of this text was common sense. You see, mathematics is a free science, a body of knowledge that anyone can derive on its own. This book is just a compendium of the works of many mathematicians through the centuries, making it possible for humankind to get to where it is today. And as life goes on, it is probable that further developments, researches and advances will continue. That was one of the main ideas in the background while writing this text -- to make the reader able to fully understand as much knowledge in linear algebra as possible, especially those concerning methods and algorithms that arose in the twentieth century.

Just as a brief comment, the first chapter of this book is dedicated to making certain topics clear to avoid confusion, and answering questions that usually arise during the first course in linear algebra. The rest of the chapters are more advanced and developed.

Author

CHAPTER 1

VECTORS: AN APPROACH TO REPRESENT SPACES

"The scientist does not study nature because it is useful; he studies it because he delights in it, and he delights in it because it is beautiful."

Henri Poincare

INTRODUCTION

Today, mathematics and physics are the main two sciences that describe nature and the world around us -- from movement to forces, from quantities to areas and volumes. In order to accomplish this, mathematics serves physics since many representations of phenomena can be well simulated by a logical mathematical approach. One of these approaches is algebra, which is probably the most used, not only because of its simplicity but also because of its great strength in complex problems.

For this to be done, a problem must be restrained first and well proposed with its variables, thereby creating the space in which the problem lives. This is a simple and philosophical statement, but many philosophical approaches have ended in robust mathematical formulations.

HISTORY OF VECTORS

As a final comment, it is relevant to talk about the history of vectors. Vectors are entities that have been studied and developed through centuries by many mathematicians. It all started in 1545 with Jerome Cardan when he published the idea of a complex number. His idea came with the following problem: "If someone says to you, divide 10 into two parts, one of which when multiplied into the other shall produce 30 or 40." It is evident that this case or question is impossible… Nevertheless, we shall solve it in this "fashion" and come up with the numbers $5+\sqrt{-15}$ and $5-\sqrt{-15}$ and saying, "putting aside the mental tortures involved, multiply $5+\sqrt{-15}$ by $5-\sqrt{-15}$, thus making $25-(-15)=25+15=40$. Hence this product is 40.

As we may see, it was difficult at first for complex numbers to be accepted. It took more than two centuries, during which many authors protested the use of these strange creations. In 1687, Newton published his "*Principia Mathematica*" in which he laid the idea of the parallelogram of forces. His statement was: "A body acted on by two forces simultaneously, will describe the diagonal of a parallelogram in the same time as it would describe the sides by those forces separately." Not vectors, but indeed really close.

In 1799, Caspar Wessel laid for the first time the geometrical representation of complex numbers. His idea was not very attractive at first, but it was best known a century later. It is important to say that Gauss also worked out the geometrical interpretation on complex numbers.

In 1813, Jean Robert Argand published his work in an attempt to find comparable methods for the analysis of complex numbers in three dimensional space, continuing the discoveries of Wessel and Gauss.

In 1843, William Rowan Hamilton, after searching for almost 13 years, found and described a way to represent complex numbers in 3D space. The history goes that he was walking with his wife at a bridge, and suddenly, after many years researching and thinking, he was enlighten with the idea of quaternions. He quickly ran and wrote in a bridge's block the rules that govern quaternions. An interesting fact is that there is a plate memorizing the moment at the place where everything happened.

Here as he walked by
on the 16th of October 1843
Sir William Rowan Hamilton
in a flash of genius discovered
the fundamental formula for
quanternion multiplication
$i^2 = j^2 = k^2 = ijk = \sim 1$
& cut it ona stone of this bridge

Figure 1: Quaternions discovery.

Source: arkadiusz-jadczyk.eu

Even though Hamilton was very much acknowledged by the mathematical community, in 1840 another mathematician, Grassmann, founded the basics of vector analysis and interpretation, giving precursors to the cross and dot products. Even though this was far beyond his time, Grassmann was ignored for almost 30 years, his discoveries where neglected and even was, indirectly, plagiarized by Cauchy. Grassmann died in 1877, after a few years his work started to get more noticed and supported.

Going back to quaternions, Hamilton had found a successor on his journey through quaternions, Peter Guthrie Tait. Tait was a math professor and was specially attracted to quaternions for their physical properties and how they could be used in other sciences, having special interest in the Nabla operator. Tait even wrote about this with William Thomson (better known as Lord Kelvin) who said that he and Tait "had a thirty-eight-year war over quaternions". Even though there was no progress with Lord Kelvin, he did have more profess with his friend James Clerk Maxwell around 1867.

As we all know, James Clerk Maxwell was the father of electromagnetism, and stated that: *"The invention of the calculus of Quaternions is a step towards the knowledge of quantities related to space which can only be compared for its importance, with the invention of triple coordinates by Descartes. The ideas of this calculus, as distinguished from its operations and symbols, are fitted to be of the greatest use in all sorts of science"*. This happened around 1873. However, after this, many authors proceeded from their own experience and rearranged the quaternionic system into a less refined system of vector analysis.

All of this continued until 1881 when Josiah Willard Gibbs, printed the first half of his *Elements of Vector Analysis*, which presents what is essentially the modern system of vector analysis. Gibbs started his analysis by reading Maxwell's *Treatise on Electricity and Magnetism*, where he was convinced to start studying quaternions and master them. He was also convinced to separate the vector part and scalar part of the product of two quaternions (which symbolizes the dot and cross product of two vectors). Later on, he discovered that he was really close to Grassmann's vector analysis but negated any inspiration from him, and was glad to "shelter himself behind two distinguished names".

Gibbs died in 1903, but Oliver Heaviside continued the journey. He was also a reader of Maxwell and was convinced of Gibbs's approach, supporting it until his death in 1925. Meanwhile, a "war" started in the mathematical community and many authors fought over quaternions or Gibbs approach. Tait was one of them, and he said that "Gibbs was retarding the quaternion progress." Heaviside, on the other hand, was more polite and wrote that "vector analysis could have been found by any mathematician by carefully examining the mechanics of the Cartesian mathematics, but to find quaternions required a genius."

After this struggle, the scientific community was worried about which system would dominate -- the system of Hamilton, Grassmann or Gibbs-Heaviside? This question was answered over the years where the Gibbs-Heaviside system was far more popular than the other ones as it is now.

We may learn many things from history, even linear algebra, since in history resides the facts of human evolution. Vectors and vector analysis is an incredible tool for many purposes and fields -- from mathematics to physics, from computers to medicine -- but it is a tool that started centuries ago with a representation of imaginary numbers, a nonsense in that time. Since then, it started to grow with the common effort of many mathematicians along the way, a process that still remains until today.

VECTOR SPACES

Before we start talking about vector spaces, it's important to talk about vectors and spaces and how to combine them. From ancient knowledge and experience, we may define space as the region in which objects exist. However, the idea of space is not limited to physical objects. In a wider sense, space is a region, physical or abstract, in which an entity exists and may be represented. This implies that ideas and certain representations of our mind also live in a given space with its own laws and rules. Every idea, thought, object, and entity lives in a certain space. However, in the light of linear algebra, we define **space** as a region where we can express entities as vectors, points, functions, curves, potential functions, sub-regions, areas and volumes.

It is important to express the idea of space to understand what a vector is. A vector usually has definitions according to its use. For example, physicists and engineers think of vectors as arrows that has a given module, direction and orientation. This is because in their fields they can express and understand quantities like forces, displacements, velocities, accelerations, potential fields, mass flows and other phenomena as vectors, and so they give it a physical meaning. On the other hand, programmers think of vectors as arrangements of data which they implement to solve system of equations, get points in space and, through a lot of interpretation by computers, transform those arrangements and list of values or "vectors" as images, videos, programs, games, or other "vectors". This is because it's useful for them to arrange data as lists. Obviously, those who have programmed before will find it easy to understand it.

In a more general sense, since a vector can be used to represent forces, lists of data, etc., a **vector** is an abstract mathematical entity that serves to represent a certain phenomenon. It's important to remark that we have not said that "a vector is ..." until the past sentence since a vector can only be described as an abstract entity that shapes itself, according to the insight of the user, to describe a situation. However, vectors have their own rules and properties, which need to be respected for them to work reasonably. In linear algebra, those rules are called **axioms** which will be explained further in this chapter.

Now that we have explained vectors and space, we can define what a vector space is. A **vector space** is a part of space, therefore a region, which can be described and represented by a certain finite, group of vectors. These vector spaces are the boundaries to certain models, like for example, two-dimensional problems. The condition of making it two-dimensional makes the problem live on a 2D vector space. A three-dimensional problem is also another vector space, and so are "n" dimensional problems and polynomials. Later on, we will define a vector space and show how to express them. Finally, we can show the axioms that rule vectors and vector spaces. These are:

*Given three vectors **u, v** and **w** which belong to a certain space V, each of them obeys the following properties and identities:*

- Commutative Property: Vectors can be summed independent of their order, and their sum also belongs to V.

  $\vec{u}+\vec{v}=\vec{v}+\vec{u} \rightarrow \in V$

- Additive identity: V has a zero vector such that:

  $\vec{u}+\vec{0}=\vec{u} \rightarrow \in V$

- Additive inverse: V has a vector denoted by -u such that:

  $\vec{u}+\overrightarrow{(-u)}=\vec{0} \rightarrow \in V$

- Scalar multiplication: Any vector multiplied or scaled by a given constant "c" will remain on V:

  $c \cdot \vec{u} \rightarrow \in V$

- Distributive property: Any group of vectors multiplied by a given group of scalars will fulfill that:

$$c \cdot (\vec{u}+\vec{v})=c \cdot \vec{u}+c \cdot \vec{v}$$

$$(c+d) \cdot \vec{u}=c \cdot \vec{u}+d \cdot \vec{u}$$

$$c \cdot (d \cdot \vec{u})=(c \cdot d) \cdot \vec{u}$$

Scalar identity: There exists a unit such that:

$$1(\vec{u}) = \vec{u}$$

Having said this, we can now proceed to the next concept of linear algebra, which is linear mapping.

LINEAR MAPPING

A **linear map**, also called a linear transformation, is a given set of algorithms that transform an entry vector into another output vector, accomplishing the axioms stated above. A linear map is a concept similar to functions -- it takes an entry and spits out an output. For example, let us suppose we have a certain vector ***u*** that we want to transform into another vector ***v*.** The way to do this is to transform them using a matrix ***A***, like this:

$$A\vec{u} = \vec{v} = L(\vec{u})$$

Where ***L(u)*** is the notation of the transformation of vector ***u***. An example of this is given below. According to the previous expression, let us suppose we have the following matrix **A** and vector ***u***, then vector ***v*** will be:

$$A = \begin{pmatrix} 1 & 2 \\ 3 & 4 \end{pmatrix}; \vec{u} = \begin{pmatrix} 1 \\ 5 \end{pmatrix} \rightarrow A\vec{u} = \begin{pmatrix} 1 & 2 \\ 3 & 4 \end{pmatrix} \cdot \begin{pmatrix} 1 \\ 5 \end{pmatrix} = \begin{pmatrix} 11 \\ 23 \end{pmatrix} = \vec{v}$$

This is a simple transformation of a single vector but we can also transform groups of vectors instead of doing it one by one. Now, let us suppose we have vectors ***a*** and ***b*** and we want to transform them into vectors ***c*** and ***d*,** respectively. With the same transformation as before:

$$A = \begin{pmatrix} 1 & 2 \\ 3 & 4 \end{pmatrix}; \vec{a} = \begin{pmatrix} 1 \\ 4 \end{pmatrix}; \vec{b} = \begin{pmatrix} -2 \\ 7 \end{pmatrix} \rightarrow A(\vec{a}\vec{b}) = \begin{pmatrix} 1 & 2 \\ 3 & 4 \end{pmatrix} \cdot \begin{pmatrix} 1 & -2 \\ 4 & 7 \end{pmatrix} = \begin{pmatrix} 9 & 12 \\ 19 & 22 \end{pmatrix}$$

Therefore,

$$L(\vec{a}) = \vec{c} = \begin{pmatrix} 9 \\ 19 \end{pmatrix}; L(\vec{b}) = \vec{d} = \begin{pmatrix} 12 \\ 22 \end{pmatrix}$$

It is important to say that what makes the transformation linear (visually speaking) is that all lines and vectors in a given space that were originally parallel will remain parallel and will be equidistant from each other after the transformation. These are the geometrical representations of the axioms stated before.

Thus, if I can input a vector ***u*** and get an output vector ***v*** using a transformation, then I can also find another transformation such that when I input vector ***v*** I will get an output vector ***u***. In other words this:

$$\vec{u} = A^{-1}\vec{v}$$

where $\mathbf{A}^{-1}$ is the inverse matrix of **A**. This is what is called the inverse transformation, which is an algorithm that reverts the original transformation and returns the space and all of its vectors back to its original state. However, finding this transformation is not always possible and we'll explain why later on.

BASIS OF A VECTOR SPACE

First, we need to discuss the basis of a space. The basis of a space is the smallest set of vectors that, when scaled and summed together, can represent all the possible vectors of the mentioned space. For example, in a one-dimensional space, like the number line, only one vector is needed to represent all the possible values on that space (since you can only move right or left on the number line). In a two-dimensional space, like the xy plane of a Cartesian set of axis, we only need two vectors to represent all the possible vectors on that plane (since you can only move right or left and up or down). Similarly, in a three-dimensional space, we need three vectors. This is related to the degrees of freedom of variation at the given space. For example, let us say we have a generic vector ***u*** in R2 with components (x, y). That vector may be represented or built with two different basis as shown below:

$$\hat{i} = \begin{pmatrix} 1 \\ 0 \end{pmatrix}; \hat{j} = \begin{pmatrix} 0 \\ 1 \end{pmatrix} \rightarrow \vec{u} = \begin{pmatrix} x \\ y \end{pmatrix} = x\begin{pmatrix} 1 \\ 0 \end{pmatrix} + y\begin{pmatrix} 0 \\ 1 \end{pmatrix} = x\hat{i} + y\hat{j}$$

$$\vec{v} = \begin{pmatrix} 1 \\ 2 \end{pmatrix}; \vec{w} = \begin{pmatrix} -1 \\ 3 \end{pmatrix} \rightarrow \vec{u} = \begin{pmatrix} x \\ y \end{pmatrix} = \left(\frac{3x+y}{5}\right)\cdot\begin{pmatrix} 1 \\ 2 \end{pmatrix} + \left(\frac{y-2x}{5}\right)\cdot\begin{pmatrix} -1 \\ 3 \end{pmatrix} = \frac{3x+y}{5}\cdot\vec{v} + \frac{y-2x}{5}\cdot\vec{w}$$

In the first example, we showed the generic vector ***u*** with the canonical basis. First, we scale the two vectors (multiplying them by a scalar) and then we sum them up, therefore obtaining ***u***. In the second example, we showed an arbitrary basis formed by vectors ***v*** and ***w***. Following the same procedure, we scale them by the factors shown and then we sum them obtaining the same vector ***u***. We just expressed the same vector using two different bases.

It is true that the basis of space may be represented by many vectors. For example, in R2 the traditional basis is the set formed by the canonical vectors ***i*** and ***j***, but (almost) any other couple of vectors may also represent the same two-dimensional space. It is important to say "almost" since not all combinations of vectors satisfy this condition. For example, if two vectors are parallel (that is, they have the same direction but not necessarily the same module and orientation) they can only represent a one-dimensional space when they are put together, since it does not matter how they are scaled and summed (in other words, how they are linearly combined). Their result will always have the same direction as theirs, and therefore, can only represent all the possible vectors on a line. Let us suppose we have the next couple of vectors:

$$\vec{a} = \begin{pmatrix} 1 \\ 2 \end{pmatrix}; \vec{b} = \begin{pmatrix} -2 \\ -4 \end{pmatrix}$$

which have the same direction, but not the same module nor the same orientation. Let us scale them by factors c_1 and c_2, respectively and then take their sum. The result is:

$$c_1 \cdot \vec{a} + c_2 \cdot \vec{b} = c_1 \cdot \begin{pmatrix} 1 \\ 2 \end{pmatrix} + c_2 \cdot \begin{pmatrix} -2 \\ -4 \end{pmatrix} = (c_1 - 2c_2) \cdot \begin{pmatrix} 1 \\ 2 \end{pmatrix} = (c_1 - 2c_2) \cdot \vec{a}$$

The resulting vector has the same direction as vector ***a***, and its module and orientation will depend on the constants c_1 and c_2. Therefore the set of vectors ***a*** and ***b*** will only be able to represent all vectors with their same direction.

The situation explained above is called ***linear dependency***. In Algebra, linear dependency is a situation in which a given vector, of a certain set of vectors, may be represented as the sum of the other vectors in that set, or better said, as a linear combination of the other vectors in that set. For example, vector ***b*** is -2 times vector ***a.*** In the case that a set of vectors is linearly dependent, then there are more vectors than the least necessary to represent the desired space. It is necessary to understand the idea that a vector is not linearly dependent on its own; it is only linearly dependent with respect to a given set of vectors (because it can be represented as the sum of the other vectors in that set).

Given a set of vectors $\boldsymbol{a_1}, \boldsymbol{a_2}, \boldsymbol{a_3}, \ldots, \boldsymbol{a_n}$ and a set of scalars $\boldsymbol{c_1}, \boldsymbol{c_2}, \boldsymbol{c_3}, \ldots, \boldsymbol{c_n}$, the set of vectors is linearly dependent if, and only if there exists a non-zero constant $\boldsymbol{c_i}$ in the solution of the following equation:

$$c_1\overrightarrow{a_1} + c_2\overrightarrow{a_2} + c_3\overrightarrow{a_3} + \ldots + c_n\overrightarrow{a_n} = \vec{0}$$

In the contrary, if all scalars are zero, then the set is called ***linearly independent***. This definition is important because the possibility of the existence of the inverse transformation matrix mentioned before relies on it.

For example, let us suppose we have three vectors ***u***, ***v*** and ***w*** with their components. This set is linearly dependent since we can express one vector as a linear combination of the other two.

$$\vec{u} = \begin{pmatrix} 1 \\ 2 \end{pmatrix}; \vec{v} = \begin{pmatrix} 3 \\ 4 \end{pmatrix}; \vec{w} = \begin{pmatrix} 2 \\ 5 \end{pmatrix}$$

$$\vec{w} = \frac{7}{2}\cdot\vec{u} + \left(\frac{-1}{2}\right)\cdot\vec{v}$$

$$\vec{v} = 7\cdot\vec{u} + (-2)\cdot\vec{w}$$

$$\vec{u} = \frac{1}{7}\cdot\vec{v} + \frac{2}{7}\cdot\vec{w}$$

As we may see, we can express any vector of the set as a linear combination of the other two. If we would only have two vectors in the set, it does not matter whether it is ***u*** and ***v***, ***u*** and ***w*** or ***v*** and ***w***, as the set would be linearly independent.

In the case when transforming a linearly independent basis of a vector space, like the canonical vectors in R2 and R3, the resulting set of vectors is linearly dependent. Thus, the transformation matrix does not have an inverse. This is because the set is linearly dependent and the transformation can take two different vectors and turn them into the same output. For example, we can show the following:

$$A=\begin{pmatrix}1&2\\2&4\end{pmatrix};\vec{u}=\begin{pmatrix}5\\0\end{pmatrix};\vec{v}=\begin{pmatrix}-1\\3\end{pmatrix}\rightarrow A\vec{u}=\begin{pmatrix}1&2\\2&4\end{pmatrix}\begin{pmatrix}5\\0\end{pmatrix}=\begin{pmatrix}5\\10\end{pmatrix}=A\vec{v}=\begin{pmatrix}1&2\\2&4\end{pmatrix}\begin{pmatrix}-1\\3\end{pmatrix}=\begin{pmatrix}5\\10\end{pmatrix}$$

As we see, the transformation can take two different linearly independent vectors and turn them into the same vector, therefore matrix ***A*** does not have an inverse. Visually speaking, this can be seen as if this transformation would take all of the two-dimensional space and shrink it into a single line making many output vectors overlap. This concept will be better understood later.

DETERMINANTS

To check if any transformation has an inverse, we use a tool called the determinant. The **determinant** of a matrix, in a two-dimensional space, is a relationship or a ratio between the area enclosed by the transformed vectors and the area enclosed by the original vectors. In a 3 dimensional space, it is a relationship among the volumes. For example, let us suppose we have a 2D space, and our matrix transformation looks like this -- a generic transformation:

$$A=\begin{pmatrix}a&b\\c&d\end{pmatrix}$$

This means that if, for example, we want to transform our canonical basis vectors ***i*** and ***j,*** then:

$$L(\hat{i})=A\hat{i}=\begin{pmatrix}a&b\\c&d\end{pmatrix}\cdot\begin{pmatrix}1\\0\end{pmatrix}=\begin{pmatrix}a\\c\end{pmatrix}$$

$$L(\hat{j})=A\hat{j}=\begin{pmatrix}a&b\\c&d\end{pmatrix}\cdot\begin{pmatrix}0\\1\end{pmatrix}=\begin{pmatrix}b\\d\end{pmatrix}$$

The determinant of matrix ***A*** would give us a relationship between the areas inscribed by the new vectors and the original vectors (the area enclosed by the vector ***i*** and ***j*** is 1 since they are both unit vectors). Geometrically, it looks like this:

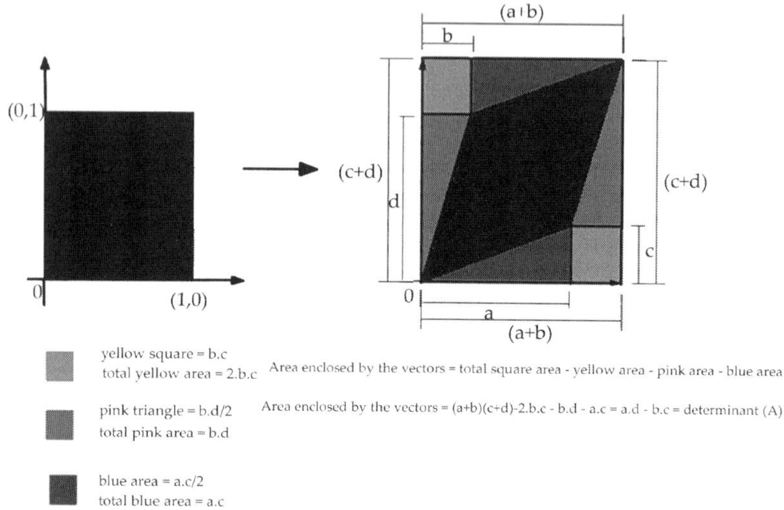

Figure 2: Determinant of a 2D transformation.

Geometrical approximations to determinants are a nice way to understand what they mean. Determinants not only limit themselves to two- or three-dimensional cases. In higher dimensions they also represent the distortion of the volume of an n-dimensional cube, turning it into an n-dimensional parallelepiped. But returning to our 2 dimensional example, following the above image, the determinant is the area of the parallelogram (***right***) divided by the area of the square (***left***), but because the area of the square is 1, then the determinant is just the area of the parallelogram.

Now that we know what a determinant is, we can develop a set of rules on how to calculate them. Many methods exist to compute them (Sarrus Rule, Laplace Method), but what we're going to talk about here is their properties. For example, let us suppose we have a certain vector space with a given basis and we apply a transformation ***A*** to it, then we apply another transformation ***B***. In other words,

$newbasis = B(A.originalbasis)$

The way to compute the determinant of the total transformation is easy in this case. If ***A*** scales the original space by its determinant and ***B*** then scales that by its determinant, then the total scale is the determinant of ***A*** times the determinant of ***B***, so:

$$det(A.B) = det(A).det(B)$$

Also, we can show a relation with the determinant of an inverse matrix. If a matrix ***A*** scales an area by its determinant, then the inverse matrix needs to revert that dividing the new area by the determinant of ***A***, so:

$$det(A^{-1}) = \frac{1}{det(A)}$$

It has not been said yet, but the determinant of a matrix can be ***negative***. It may sound strange to scale an area by a negative value by it has something to do with orientation rather than with size. For example, in the regular xy plane, vector ***i*** is always at the right side of vector ***j***. If we make a transformation and the new vector ***L(i)*** is at the left side of the new vector ***L(j)*** then the determinant is negative, but an area in the vector space will be equally scaled by the absolute value of the determinant. Actually, this property is interesting, because if we swap two columns in a matrix, the determinant will be -1 times the determinant of the new matrix, because we have changed the orientation of the resulting vectors.

Now that we have reached this point, we can say that if a matrix transforms a linearly independent basis of a vector space into a linearly dependent new set of vectors, then the matrix transformation does not have an inverse. This means that the transformation can take two different vectors and transform them into the same vector. Meaning, there is no one-to-one correspondence. The only way for this to happen is that the transformation takes all vectors (for example, in an R3) and overlaps them into a plane or a line, making infinite vectors have the same output. Geometrically speaking, this means that if we have a basis in R3 (meaning we can represent a unit volume), apply a transformation and get a linearly dependent set of vectors, then we would not be able to represent another volume (because two or more vectors have the same direction). We would only be able to represent an area or a length, meaning no volume, and therefore zero determinant.

If we would be talking in a two-dimensional space and the new set of vectors are linearly dependent, then they have the same direction. Therefore, they can only express a line and there is no area between them. The same happens in an n-dimensional space -- if the new set of vectors is linearly dependent, then the new set would only express a lower dimension volume, and so the determinant would be zero. In all of these cases, the determinant of the transformation matrix is zero because the linear dependency of the vectors reduces the dimension of the vector space.

DIMENSION OF A VECTOR SPACE

It would be important to talk about a concept that is useful in these situations. This is the dimension of a vector space. The **dimension of a vector space** is the smallest number of vectors in a basis that can express the mentioned space. For example, the dimension in R2 is 2, since with a linear combination of two linearly independent vectors we can get any vector in R2. The same is true with R3 -- the dimension is 3. In the number line, the dimension is 1.

What actually happens when a transformation matrix has a zero determinant is that we are reducing the dimension of the vector space. For example, in R2 (two-dimensional vector space) we could reduce that space into a line (one-dimensional vector space) or into a point (zero-dimensional vector space). The same thing happens in R3. But what is interesting is that that when we apply transformations we can only remain in our dimensional space or move to a lower dimensional space, not to a higher one.

A question arises when the transformation matrices are not square but rectangular. For example, we have a non-square matrix $\boldsymbol{A}$, and two linearly independent vectors $\boldsymbol{u}$ and $\boldsymbol{v}$ that are a basis of R2. If we transform them to get to R3, then

$$A = \begin{pmatrix} 1 & 2 \\ 3 & 4 \\ 5 & 6 \end{pmatrix}; \vec{u} = \begin{pmatrix} -1 \\ 5 \end{pmatrix}; \vec{v} = \begin{pmatrix} 2 \\ -7 \end{pmatrix} \rightarrow A(\vec{u}\vec{v}) = \begin{pmatrix} 1 & 2 \\ 3 & 4 \\ 5 & 6 \end{pmatrix} \begin{pmatrix} -1 & 2 \\ 5 & -7 \end{pmatrix} = \begin{pmatrix} 9 & -12 \\ 17 & -22 \\ 25 & -32 \end{pmatrix}$$

We only get two vectors even though they are in R3. It's because even if they have three components, they are not enough to express the whole 3D space. They can only express a plane, and therefore remains in a two-dimensional space even when the vectors have three components.

These are the basics of vector spaces. However, there are some more definitions and important procedures that are worthy of mention. For example, we just showed that it is impossible in a transformation to increase the dimension of the original space; but what happens if a transformation reduces the dimension? By how much is the dimension reduced? That is what we are going to discuss now.

REDUCING THE DIMENSION

In case a transformation reduces the dimension of a vector space, for example, from R3 to R2, then the new basis cannot express all the original space. If a transformation from R3 produces a set of vectors that can only express a plane, then the transformation will leave behind an enormous quantity of vectors which cannot be created by the new basis, literally leaving a whole dimension. In order to calculate the possible vectors that we can create with the new basis we perform the following analysis.

We first take the new set of vectors and make a generic linear combination of them, then find restrictions in that linear combination. Let us keep that for a moment and give an example. Given the vectors ***u***, ***v*** and ***w***, which are the bases of a transformation of a vector space, show the vector space which they can generate:

$$\vec{u} = \begin{pmatrix} 1 \\ 2 \\ 3 \end{pmatrix}; \vec{v} = \begin{pmatrix} 2 \\ 3 \\ 5 \end{pmatrix}; \vec{w} = \begin{pmatrix} 1 \\ 5 \\ 6 \end{pmatrix}$$

First, we make a linear combination of them and equate it to a generic vector of components a, b and c.

$$c_1\vec{u}+c_2\vec{v}+c_3\vec{w}=c_1\cdot\begin{pmatrix}1\\2\\3\end{pmatrix}+c_2\cdot\begin{pmatrix}2\\3\\5\end{pmatrix}+c_3\cdot\begin{pmatrix}1\\5\\6\end{pmatrix}=\begin{pmatrix}a\\b\\c\end{pmatrix}$$

To make it easier, we can express it in matrix notation and apply Gaussian elimination:

$$\begin{pmatrix}1&2&1\\2&3&5\\3&5&6\end{pmatrix}\begin{pmatrix}c_1\\c_2\\c_3\end{pmatrix}=\begin{pmatrix}a\\b\\c\end{pmatrix}\rightarrow\begin{matrix}R_2=R_2-2R_1\\R_3=R_3-3R_1\end{matrix}$$

$$\begin{pmatrix}1&2&1\\0&-1&3\\0&-1&3\end{pmatrix}\begin{pmatrix}c_1\\c_2\\c_3\end{pmatrix}=\begin{pmatrix}a\\b-2a\\c-3a\end{pmatrix}\rightarrow R_3=R_3-R_2$$

And as we see

$$\begin{pmatrix}1&2&1\\0&-1&3\\0&0&0\end{pmatrix}\begin{pmatrix}c_1\\c_2\\c_3\end{pmatrix}=\begin{pmatrix}a\\b-2a\\c-a-b\end{pmatrix}$$

The former procedure is a way of solving a system of equations, which is something I assume you already know. We see that the final equation is a row of zeros equal to c – a - b. Therefore, we have a restriction which is:

$$c-a-b=0\rightarrow c=a+b$$

This is our restriction that we must observe, and the set of vectors given before can only generate vectors which have the above restriction. From here we can calculate another basis of this vector space:

$$\begin{pmatrix}a\\b\\c\end{pmatrix}=\begin{pmatrix}a\\b\\a+b\end{pmatrix}=\begin{pmatrix}a\\0\\a\end{pmatrix}+\begin{pmatrix}0\\b\\b\end{pmatrix}=a\cdot\begin{pmatrix}1\\0\\1\end{pmatrix}+b\cdot\begin{pmatrix}0\\1\\1\end{pmatrix}$$

This is a general approach to check the dimension of a vector space, which is 2 in this case. Therefore, the set of vectors generates a plane. It is to be noted that the basis given at the beginning was linearly dependent. If the vectors have been linearly independent, then the resulting basis would have had three vectors, the dimension would be 3, and we would not have any restriction. Restrictions only appear when the set of vectors are linearly dependent. And as restrictions grow in number, the dimension of the resulting space gets lower. If we have started the problem with any group of two vectors ***u***, ***v*** and ***w***, we would have gotten the same result. That is, this procedure involves removing the unnecessary vectors, finding a restriction, and returning a linearly independent basis, which in the above example are vectors (1, 0, 1) and (0, 1, 1).

Also, the vector space which we will call ***H***, can be expressed like this:

$$H = \left[\begin{pmatrix} a \\ b \\ c \end{pmatrix} \in R^3 \rightarrow c - b - a = 0 \right]$$

ORTHOGONAL COMPLEMENT

Think about this: if a transformation takes vectors from a vector space of dimension 3 to dimension 2, then where did the other dimension go? If ***H*** is a vector space of $\boldsymbol{R^n}$ with dimension ***m*** lower than ***n***, then what are the "missing" dimensions? The answer to this question is called the orthogonal complement of ***H***. In few words, having a vector space ***H*** in $\boldsymbol{R^n}$ with dimension ***m*** lower than ***n***, the **orthogonal complement** of ***H*** represents another vector space whose vectors are orthogonal to all vectors in ***H***, that is, it represents the missing dimensions of ***H*** to represent all vectors in $\boldsymbol{R^n}$. It is important to say that if ***H*** is a vector space in $\boldsymbol{R^n}$, then the dimension of ***H*** plus the dimension of its orthogonal complement is equal to n. Thus,

$$orthogonalcomplementofH = H^{\perp}$$

$$H \wedge H^{\perp} \in R^n \rightarrow dimensionofH + dimensionofH^{\perp} = n$$

In the former example, the vectors have three components because they are in R3, but since ***H*** has a dimension equal to two, then the orthogonal complement has a dimension equal to one.

The orthogonal complement of a vector space H is calculated taking a generic vector ***t*** with components (x, y, z), and applying dot product equal to 0 with the basis vectors of H. It is assumed that you already know what a dot product is. However, a proof of it in R3 will be demonstrated:

SHORT PARENTHESIS: THE DOT PRODUCT

Let us suppose we have three vectors as shown in the image below, vector ***A***, vector ***B*** and vector ***B-A***. They all have a given length or module, and vector ***A*** and ***B*** form an angle theta. Let us suppose ***A*** and ***B*** have components:

$$\vec{A}=\begin{pmatrix}a_1\\a_2\\a_3\end{pmatrix};\vec{B}=\begin{pmatrix}b_1\\b_2\\b_3\end{pmatrix};\left(\vec{B}-\vec{A}\right)=\begin{pmatrix}b_1-a_1\\b_2-a_2\\b_3-a_3\end{pmatrix}$$

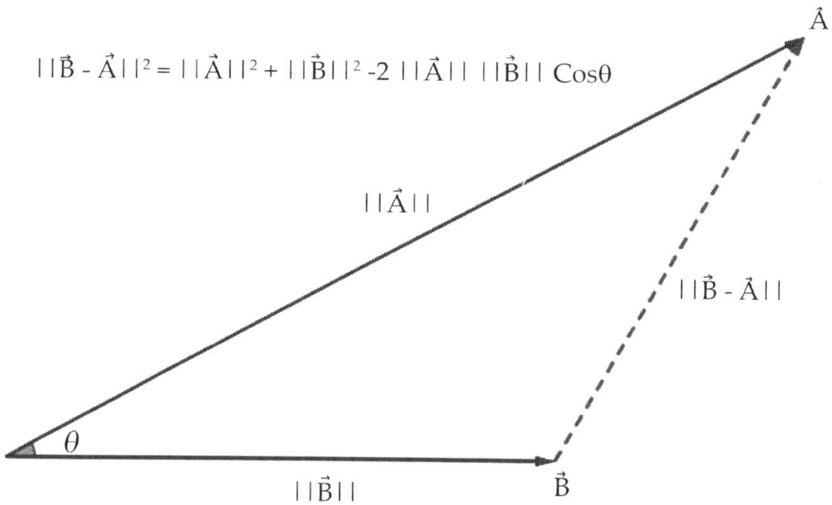

Figure 3: Law of cosines.

Source: wiki.roblox.com

We know that the module of a vector is the square root of the sum of the squares of its components. In a vector of n components, the module is given by

$$\| \vec{a} \| = \sqrt{\sum_{i=1}^{i=n} (a_i)^2}$$

Following the law of cosines seen in the picture above and introducing the module of the vectors according to their components, we get

$$\|(\vec{B}-\vec{A})\|^2=\|\vec{A}\|^2+\|\vec{B}\|^2-2\|\vec{A}\|\ \|\vec{B}\|\cos(\theta)$$

$$\left(\sqrt{(b_1-a_1)^2+(b_2-a_2)^2+(b_3-a_3)^2}\right)^2=\left(\sqrt{(a_1^2+a_2^2+a_3^2}\right)^2+\sqrt{(b_1^2+b_2^2+b_3^2})^2-2\cdot\sqrt{a_1^2+a_2^2+a_3^2}\cdot\sqrt{b_1^2+b_2^2+b_3^2}\cdot\cos(\theta)$$

The squares on the modules take the square roots, leaving the equation easier to manipulate. Then we develop the squares and get

$$\left(b_1^2+b_2^2+b_3^2\right)+\left(a_1^2+a_2^2+a_3^2\right)-2\left(a_1b_1+a_2b_2+a_3b_3\right)=\left(a_1^2+a_2^2+a_3^2\right)+\left(b_1^2+b_2^2+b_3^2\right)-2\cdot\sqrt{a_1^2+a_2^2+a_3^2}\cdot\sqrt{b_1^2+b_2^2+b_3^2}\cdot\cos(\theta)$$

.

$$a_1b_1+a_2b_2+a_3b_3=\sqrt{a_1^2+a_2^2+a_3^2}\cdot\sqrt{b_1^2+b_2^2+b_3^2}\cdot\cos(\theta)$$

We will call the left side of the equation the dot product between ***A*** and ***B***, and we will represent the right side with the modules of the vectors:

$$\vec{A}\cdot\vec{B}=\|\vec{A}\|\ \|\vec{B}\|\cos(\theta)$$

As we see, if vectors A and B are orthogonal, meaning the angle between them is 90 degrees, then their dot product is zero. In case you are wondering what the dot product geometrically means, suppose B is a unit vector (therefore, its magnitude is 1). Thus,

$$\vec{A}\cdot\hat{B}=\|\vec{A}\|\cos(\theta)$$

This is the projection of ***A*** in the direction of ***B***. This can be seen in the above image, since the module of ***A*** times the cosine of theta is the length of ***A*** projected into the direction of ***B***. The dot product can also tell us if two vectors have the same orientation or not. For example, if ***A*** and ***B*** have the same orientation, then theta is between 0 and 90 degrees, and the dot product is positive; but if ***A*** and ***B*** have opposite orientation, then theta is between 90 and 180 degrees, and the dot product is negative. Something important to add is that theta is the smallest angle between vectors ***A*** and ***B*****,** therefore, it is between 0 and 180 degrees. A really interesting fact is that

$$\vec{A}-\left(\vec{A}\cdot\hat{B}\right)\hat{B}\perp\vec{B}$$

The above expression means that vector ***A*** minus its projection over vector ***B*** is orthogonal to ***B***. This is because we are subtracting from ***A*** anything that is in the direction of ***B*** (in case ***A*** and ***B*** would have the same direction, ergo are parallel, then the result of subtracting vector ***A*** minus its projection would be zero). This procedure will be useful later with something called Gram- Schmidt orthonormalization. But for now, let us return to orthogonal complements.

Following the example on vector space H, we will now show how to calculate its orthogonal complement.

$$\vec{t} = \begin{pmatrix} x \\ y \\ z \end{pmatrix} \rightarrow \vec{t} \cdot basisofH = 0 \rightarrow \begin{pmatrix} x \\ y \\ z \end{pmatrix} \cdot \begin{pmatrix} 1 \\ 0 \\ 1 \end{pmatrix} = 0; \begin{pmatrix} x \\ y \\ z \end{pmatrix} \cdot \begin{pmatrix} 0 \\ 1 \\ 1 \end{pmatrix} = 0$$

The above dot products give equations

$$\begin{matrix} x + z0 \\ y + z = 0 \end{matrix} \rightarrow x = y = -z \rightarrow \vec{t} = \begin{pmatrix} x \\ y \\ z \end{pmatrix} = \begin{pmatrix} -z \\ -z \\ z \end{pmatrix} = z \begin{pmatrix} -1 \\ -1 \\ 1 \end{pmatrix}$$

$$\left[H^{\perp} = \begin{pmatrix} x \\ y \\ z \end{pmatrix} \in R^3 \rightarrow x = y = -z \right]$$

As we can see, the orthogonal complement has a dimension of 1 because the dimension of the vectors is 3 (vectors with 3 components) while the dimension of H is 2. In short, 3 – 2 = 1.

Now that we know this, we can proceed with something very important in linear algebra which is the Gram Schmidt orthonormalization.

GRAM SCHMIDT ORTHONORMALIZATION

The Gram Schmidt orthonormalization is a procedure which takes a set of vectors in a basis and assures that at the end all vectors shall be orthogonal to each other and will be unit vectors. These are desirable conditions in some cases since it is easier easy to work with this kind of set of vectors because as they are unit vectors (they only have a direction and their modules are not appearing anywhere in equations). Also, as they are orthogonal, their dot and cross products simplify a lot. Explaining why this conditions are desirable is not easy. It has to do more with the ease in calculations and problems approach rather than words. Just to have an idea about it, the canonical vectors of an $\mathbf{R}^n$ space are orthonormal, and a Frenet frame is also a set of orthonormal vectors.

Since we mentioned the cross product, we shall take a parenthesis here to talk about it and give a proof on how to calculate it in R3.

SHORT PARENTHESIS: THE CROSS PRODUCT

The cross product of two vectors ***u*** and ***v*** is an operation that calculates another vector that is perpendicular to ***u*** and ***v***. This may sound simple but there is much to say. As math is the universal language, let us express our problem:

Given two vectors ***u*** and ***v***, calculate a unique vector ***w*** that is perpendicular to ***u*** and ***v***. Let us say vectors ***u***, ***v*** and ***w*** have components

$$\vec{u} = \begin{pmatrix} u_1 \\ u_2 \\ u_3 \end{pmatrix}; \vec{v} = \begin{pmatrix} v_1 \\ v_2 \\ v_3 \end{pmatrix}; \vec{w} = \begin{pmatrix} w_1 \\ w_2 \\ w_3 \end{pmatrix}$$

Therefore, if we want ***w*** to be perpendicular to ***u*** and ***v***, then their dot products shall be zero, giving the equations

$$u_1 w_1 + u_2 w_2 + u_3 w_3 = 0$$

$$v_1 w_1 + v_2 w_2 + v_3 w_3 = 0$$

We have two equations with three unknowns. Therefore, we have infinite solutions, and actually yes, they are infinite vectors that are perpendicular to both ***u*** and ***v***. The only difference among them all is their module. For example, if you give a value to one of the components of ***w*** you could solve the equations, and you would get the desired vector. If you give the same component another value you would get another vector parallel to the first you found, and their only difference will be their size.

To complete the problem, we need to assign a logical value to the module of ***w***. Many interesting options come in mind. We could say that ***w*** is a unit vector (therefore, module 1), or its module would be the module of ***u*** times the module of ***v***. In other words,

$$option1 \rightarrow \| w \| = 1$$

$$option2 \rightarrow \| w \| - \| u \| \; \| v \|$$

Before we proceed, let us look at a graph of vectors u and v.

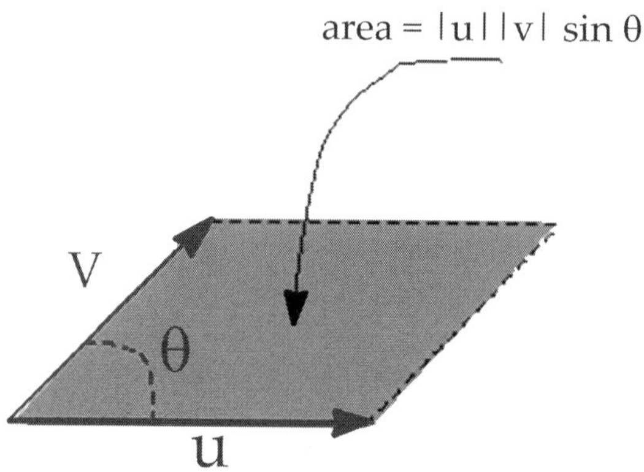

Figure 4: Area between two vectors.

Source: chortle.ccsu.edu/

Let us suppose we say that vector ***w*** has module according to option 1 or option 2. In either case, what happens if vectors ***u*** and ***v*** are parallel? Then the two dot products shown above are linearly dependent and there is no way to compute a reasonable result for ***w*.** To solve this problem, we need to assign a module that is logical even when ***u*** and ***v*** are parallel. For this, we say that the module of ***w*** is the area between ***u*** and ***v*,** such that in case they are parallel, the area is 0; and the only way for a vector to have 0 magnitude is to be the zero vector.

This may sound like something out of the sleeve, but a singularity exists when ***u*** and ***v*** are parallel, because there is no way to calculate ***w*** in that singularity. The way of solving this is by applying the concept of area between vectors, because if there is no way of solving the problem in that singularity then there is no vector, and that is the zero vector.

Now that we have the system of equations completed, we will proceed with solving for the components of ***w***:

$$\vec{u} \cdot \vec{w} = 0 \rightarrow u_1.w_1 + u_2.w_2 + u_3.w_3 = 0$$

$$\vec{v} \cdot \vec{w} = 0 \rightarrow v_1.w_1 + v_2.w_2 + v_3.w_3 = 0$$

$$\| w \| = \| u \| \ \| v \| \, sin(\theta)$$

First, let us rearrange the last equation, using trigonometry, the dot product and a little bit of algebra:

$$\|w\| = \|u\|\|v\|\sin(\theta)$$

$$\|w\|^2 = \|u\|^2\|v\|^2\sin^2(\theta)$$

$$\|w\|^2 = \|u\|^2\|v\|^2(1 - \cos^2(\theta))$$

$$\|w\|^2 = \|u\|^2\|v\|^2(1 - (\frac{\vec{u} \cdot \vec{v}}{\|u\|\|v\|})^2)$$

$$\|w\|^2 = \|u\|^2\|v\|^2 - (\vec{u} \cdot \vec{v})^2$$

$$(w_1^2 + w_2^2 + w_3^2) = (u_1^2 + u_2^2 + u_3^2) \cdot (v_1^2 + v_2^2 + v_3^2) - (u_1v_1 + u_2v_2 + u_3v_3)^2$$

$$(w_1^2 + w_2^2 + w_3^2) = (u_1v_2 - u_2v_1)^2 + (u_1v_3 - u_3v_1)^2 + (u_2v_3 - u_3v_2)^2$$

From here we can solve the system of equations, which will be left to you as an exercise. An advice for doing so is to solve for w_1 and w_2 in terms of w_3 in the first two equations, and then substitute those expressions to the last one. Nevertheless, there are many ways of solving this system of equations, and regardless of them, the end result would look like this:

$$w_1 = \pm(u_2v_3 - u_3v_2)$$
$$w_2 = \mp(u_1v_3 - u_3v_1)$$
$$w_3 = \pm(u_1v_2 - u_2v_1)$$

The plus and minus signs are of very critical relevance, since in either case, the vector satisfies all the conditions. Actually, the two solutions are the same vector with the same module and direction, only with opposite orientation. To close the problem, we only need to input another condition, and it is given by the ***right hand rule***.

The ***right hand rule*** is the convention taken took in vector spaces and algebra to choose correctly the orientation of vectors in cross products and more. In a literal sense, you could create your own space with a "left hand rule" (like when programming games) but the right hand rule is a standard convention so that all of us can understand ourselves and our equations. In making the cross product of vectors ***u*** and ***v*** and applying the right hand rule, we will get

$$\vec{u}\times\vec{v}=\begin{pmatrix} u_2v_3-u_3v_2 \\ -\left(u_1v_3-u_3v_1\right) \\ u_1v_2-u_2v_1 \end{pmatrix};\vec{v}\times\vec{u}=\begin{pmatrix} v_2u_3-v_3u_2 \\ -\left(v_1u_3-v_3u_1\right) \\ v_1u_2-v_2u_1 \end{pmatrix}\rightarrow\vec{u}\times\vec{v}=-\vec{v}\times\vec{u}$$

It's important to say that the cross product is not commutative but it does follow the relation stated above. Also, the cross product is particularly shown with a cross between vectors (thus, its name). As a final comment on the cross product, in case you are wondering, yes, the cross product exactly coincides with the following determinant:

$$\vec{w}=\begin{vmatrix} \hat{i} & \hat{j} & \hat{k} \\ u_1 & u_2 & u_3 \\ v_1 & v_2 & v_3 \end{vmatrix}$$

However, presenting the cross product as a determinant is a little strange since the determinant shows the distortion of a unit volume during a transformation, but in this case, the calculation of the above determinant and the cross product coincides perfectly. This is a plus since now we have an easy method to calculate cross products.

An important detail with the dot and cross products is that we can trick them to get very useful relations when working with vectors. For example, if the dot product of two vectors is 0 then those two vectors are orthogonal; but if the cross product of two vectors gives the zero vector, then those two vectors are parallel.

Returning to our original problem, we can continue the Gram Schmidt orthonormalization explanation. As we said before, the Gram Schmidt process takes a set of vectors and turns them into a set of unit and orthogonal vectors. This is done in a simple way -- take one arbitrary vector of the set and normalize it, then grab another vector on the set and take away its projection from the preceding vectors, then normalize it, and repeat the process with all the vectors in the basis. For example, let us do this process with the basis of vector space H we had in the previous discussion. First, we take an arbitrary vector from the basis and normalize it. Let us call the vectors on the basis, $\boldsymbol{v}_{i;}$ the vectors without the projection, $\boldsymbol{u'}_i$; and the normalized vectors, $\boldsymbol{u}_i$.

$$\vec{v_1} = \begin{pmatrix} 1 \\ 0 \\ 1 \end{pmatrix}; \rightarrow \vec{u_1} = \frac{1}{\sqrt{2}}\begin{pmatrix} 1 \\ 0 \\ 1 \end{pmatrix}$$

Now, let us grab the other vector and take away its projection from the first one.

$$\vec{v_2} = \begin{pmatrix} 0 \\ 1 \\ 1 \end{pmatrix} \rightarrow \vec{u}'_2 = \vec{v_2} - \left(\vec{v_2} \cdot \vec{u_1}\right)\vec{u_1} = \begin{pmatrix} 0 \\ 1 \\ 1 \end{pmatrix} - \frac{\left(0\cdot 1 + 1\cdot 0 + 1\cdot 1\right)}{\sqrt{2}} \cdot \frac{1}{\sqrt{2}}\begin{pmatrix} 1 \\ 0 \\ 1 \end{pmatrix}$$

$$\vec{u}'_2 = \begin{pmatrix} 0 \\ 1 \\ 1 \end{pmatrix} - \frac{1}{2}\begin{pmatrix} 1 \\ 0 \\ 1 \end{pmatrix} = \begin{pmatrix} -1/2 \\ 1 \\ 1/2 \end{pmatrix}$$

$$\vec{u_2} = \frac{\sqrt{2}}{2\sqrt{3}}\begin{pmatrix} -1 \\ 2 \\ 1 \end{pmatrix}$$

We now have two unit and orthogonal vectors $\boldsymbol{u}_1$ and $\boldsymbol{u}_2$ which form an orthonormal basis of our vector space ***H***.

CHAPTER 2

TRANSFORMATIONS AND LINEAR TRANSFORMATIONS

"The study of mathematics, like the Nile, begins in minuteness but ends in magnificence"

Charles Caleb Colton

INTRODUCTION

In the preceding chapter we focused on the subtle but very important subjects like dimensions, vector spaces, determinants and more. Those are the primary tools to understand the next stages of algebra, or at least, linear algebra. It is important to say that algebra is a very immense field of mathematics. And even though it started as a science to represent the world around us during the beginning of history, it has now evolved into a deeper level of ***abstraction***. That word is key, ***abstraction***.

For example, when we were just small children, we grew up and started to learn how to think. We learned how to count numbers and count units like our fingers, from 1 to 10. And then, we were taught, without the notion of it, the ***natural numbers***. Later on we discovered by ourselves that there are more numbers than we thought -- they are the negative numbers, which we thought as something that are "missing" since they are negative, and therefore we continued learning the ***integers*** (as a matter of fact, the set of integers is denoted by the letter Z, which comes from the German word *zahlen* which means "numbers").

Following that, some of us might have gotten into a situation of sharing a cookie with our friends, so we split it in half. That was something simple for a kid, but we actually performed a mathematical operation which is division, and encountered a new set of numbers called the ***rational numbers***. When we think of it, the rational numbers are like numbers inside or between other numbers, which is a complement of what we have learned on the past. By the way, the rational numbers are denoted by the letter Q which comes quotient.

The Greeks thought that all numbers where rational, but now we know that there exist the ***irrational Numbers***. Going further, the set of numbers made by the natural numbers, the integers, the rational and the irrational numbers make the set of the ***real numbers***. There are also even more ways to catalog numbers, like the ***transcendental numbers***, which are numbers that cannot be written as a result from an equation, but only as a constant. Transcendental numbers take their name because they behave as though they are transcending mathematics.

As you can see, looking at that particular example of the different sets of numbers, we started on a simple thing, the natural numbers, and ended up in something more complex, the real numbers. This is because it is a process of progressive abstraction. Abstraction, in mathematics, is the key to expand knowledge, since it generalizes the knowledge we had before with another one new we get from learning and researching. As we left linear transformations in the preceding chapter, let us take it from there.

TRANSFORMATIONS

As we have defined before, a **transformation** is a function that takes an input and gives an output. In the preceding chapter we focused our attention in linear transformations, but that does not mean that there does not exist other kinds of transformations. In reality any transformation is possible, like taking a number and giving its square, or taking a number and giving its absolute value. These are examples of transformations that happen in many places and other fields of mathematics, however, they are not linear.

The thing with transformations is that they change depending on the user's needs or his level of abstraction. For example, physicists require a great level of abstraction in their formulations since almost any transformation can arise from their formulations. For example, let us look at the following sum:

$$\sum_{i=1}^{\infty} i = 1 + 2 + 3 + \ldots$$

From arithmetic, we know that if we cut that sum at a finite value, let us say "n", then that sum is equal to:

$$\sum_{i=1}^{n} i = \frac{n(n+1)}{2}$$

But scientists and mathematicians believe that when that sum goes on forever until infinity, it equals to

$$\sum_{i=1}^{\infty} i = 1 + 2 + 3 + \ldots = -\frac{1}{12}$$

Which is probably surprising, since how is it possible to add an infinite number of positive integers and get a negative fraction? Well, scientist believe it to be that way, since that sum appears in many equations and it's critical in obtaining the 26 dimensions of String Theory. Finitely speaking, that does not make sense at all, but that sum is infinite, it goes on forever and we cannot think of it as a finite situation. That is a very clear example of a whole new level of abstraction and a sort of transformation.

Going back to our topic, giving examples of transformations is sort of a trivial thing since all transformations are possible. But just to give you an idea, it is possible to transform polynomials into matrices, or matrices into scalars, or vectors into equations, or points into differential equations, depending on the needs of the mathematician. In this case, for example, we will focus on certain transformations, which are linear transformations.

Linear transformations are simple but very useful and powerful because of their properties, since they only scale and add vectors. For example, let us look at the following properties: If ***T*** is a linear transformation, and ***a*** and ***b*** are vectors, then it follows that

$$T(\vec{a} + \vec{b}) = T(\vec{a}) + T(\vec{b})$$

$c \in \mathbb{R} \rightarrow T(c\vec{a}) = cT(\vec{a})$

These are consequences of the linearity of transformations. If you recall from the first chapter, we saw problems in which a transformation takes a vector from a vector space and gets it into another vector space. The following image explains that:

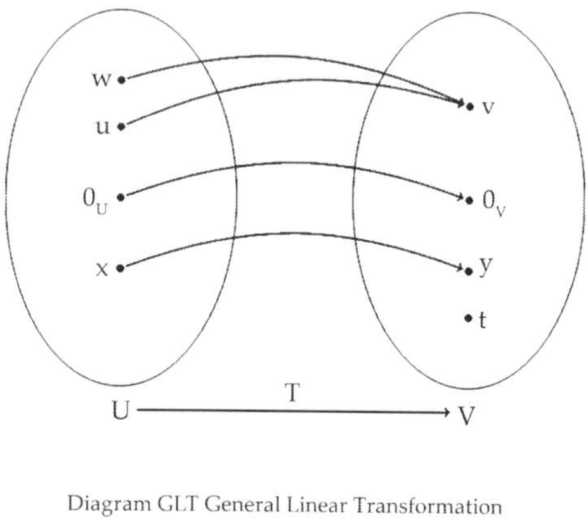

Figure 5: General linear transformation.

Source: http://linear.ups.edu

By the way, the transformation shown in the above diagram does not have an inverse (Hint: it takes two vectors from ***U*** and takes them to the same output vector in ***V***) and it shows a very important property -- every linear transformation that takes the zero vector, also gives the zero vector. Having this in mind, let us look at certain transformations to identify which is linear and which one is not, and to notice several things.

TRANSFORMATION EXAMPLES

Example 1. Given the following transformation, show that it is linear.

$$\vec{u} = \begin{pmatrix} u_1 \\ u_2 \\ u_3 \end{pmatrix} \in V \rightarrow R^3; T(\vec{u}) = (u_1 + u_2 + u_3)x^2 + (2u_2)x + u_3 \rightarrow P(x)_2$$

The above transformation takes a vector from $\boldsymbol{R}^3$ and takes it to the polynomials of second order. To determine if a transformation is linear, the first thing to do is to check if the transformation takes the zero vector to the zero element. Thus,

$$\vec{0}=\begin{pmatrix}0\\0\\0\end{pmatrix}\rightarrow T\left(\vec{0}\right)=\left(0+0+0\right)x^2+\left(2\cdot 0\right)x+0=0$$

After this, we can show the first property: *the transformation of a sum of elements is the sum of the transformations of the elements.*

$$\vec{u}=\begin{pmatrix}u_1\\u_2\\u_3\end{pmatrix};\vec{v}=\begin{pmatrix}v_1\\v_2\\v_3\end{pmatrix}\rightarrow T\left(\vec{u}+\vec{v}\right)=\left(u_1+u_2+u_3+v_1+v_2+v_3\right)x^2+2\left(u_2+v_2\right)x+\left(u_3+v_3\right)$$

$$T\left(\vec{u}+\vec{v}\right)=\left(u_1+u_2+u_3\right)x^2+\left(v_1+v_2+v_3\right)x^2+2u_2x+2v_2x+u_3+v_3=T\left(\vec{u}\right)+T\left(\vec{v}\right)$$

Then we proceed to show the next property:

$$\vec{u}=\begin{pmatrix}u_1\\u_2\\u_3\end{pmatrix}\rightarrow T\left(c\cdot\vec{u}\right)=\left(c\cdot u_1+c\cdot u_2+c\cdot u_3\right)x^2+2\left(c\cdot u_2\right)x+\left(c\cdot u_3\right)=cT\left(\vec{u}\right)$$

Therefore, ***T*** is a linear transformation.

Example 2. Let us look at the following transformation:

$$y=ax^3+bx^2+cx+d\in P\left(x\right)_3;T\left(y\right)=\begin{pmatrix}a & b\\c & d+1\end{pmatrix}\rightarrow M_{2x2}$$

This is a transformation that takes a polynomial and turns it into a 2×2 matrix. If we have the next polynomial then its transformation would be

$$y=1x^3+2x^2+3x+4;T\left(y\right)=\begin{pmatrix}1 & 2\\3 & 4+1\end{pmatrix}=\begin{pmatrix}1 & 2\\3 & 5\end{pmatrix}$$

which is a valid transformation. However, it is not linear since it does fulfill the third property of linear transformations

$$y=ax^3+bx^2+cx+d\in P\left(x\right)_3;m=constant\rightarrow T\left(m\cdot y\right)=\begin{pmatrix}m\cdot a & m\cdot b\\m\cdot c & m\cdot d+1\end{pmatrix}\neq m\cdot\begin{pmatrix}a & b\\c & d+1\end{pmatrix}$$

It is important to note that it is not linear because of the number 1 on the last term.

REAL TRANSFORMATION EXAMPLES

We just saw a couple of random examples of transformations. However, most common transformations look like the following:

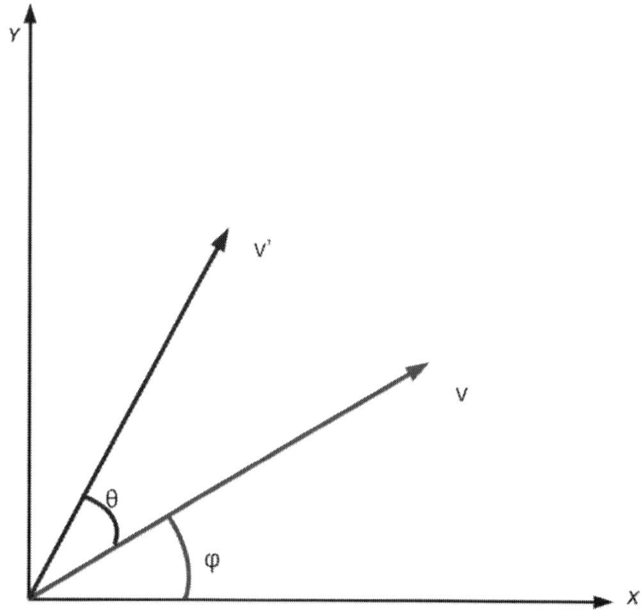

Figure 6: Vector rotation.

Vector Rotation: Let us say we have a vector in the xy plane and we know its components. And let us say we want to rotate it certain degrees and obtain its new components. This problem can be solved and deduced in several ways, but let us do it the following way: In the image shown above we have vector ***v*** (colored in red with φ degrees with respect to the x axis) and we want to rotate it θ degrees counterclockwise (ergo, we want to make it the new vector ***v'*** colored in blue), and get its new coordinates. We will take advantage of the fact that the module of ***v*** does not change during its rotation. Using trigonometry:

$$v = \begin{pmatrix} v_x \\ v_y \end{pmatrix} = |v| \begin{pmatrix} cos\varphi \\ sin\varphi \end{pmatrix}$$

$$v' = \begin{pmatrix} v'_x \\ v'_y \end{pmatrix} = |v| \begin{pmatrix} cos(\varphi+\theta) \\ sin(\varphi+\theta) \end{pmatrix} = |v| \begin{pmatrix} cos(\varphi)cos(\theta) - sin(\varphi)sin(\theta) \\ sin(\varphi)cos(\theta) + cos(\varphi)sin(\theta) \end{pmatrix}$$

$$v' = \begin{pmatrix} v_x cos(\theta) - v_y sin(\theta) \\ v_y cos(\theta) + v_x sin(\theta) \end{pmatrix} = \begin{pmatrix} \cos\theta & -\sin\theta \\ \sin\theta & \cos\theta \end{pmatrix} \cdot \begin{pmatrix} v_x \\ v_y \end{pmatrix} = \begin{pmatrix} \cos\theta & -\sin\theta \\ \sin\theta & \cos\theta \end{pmatrix} \cdot \vec{v}$$

This is a very common transformation in $\boldsymbol{R}^2$, however, it can also be done in $\boldsymbol{R}^3$ with the following transformations (note that these are the expressions for counterclockwise rotations over the axis):

$$R_z = \begin{pmatrix} \cos\theta & -\sin\theta & 0 \\ \sin\theta & \cos\theta & 0 \\ 0 & 0 & 1 \end{pmatrix}; R_x = \begin{pmatrix} 1 & 0 & 0 \\ 0 & \cos\alpha & -\sin\alpha \\ 0 & \sin\alpha & \cos\alpha \end{pmatrix}; R_y = \begin{pmatrix} \cos\beta & 0 & \sin\beta \\ 0 & 1 & 0 \\ -\sin\beta & 0 & \cos\beta \end{pmatrix}$$

We can see that the matrix transformations are really simple and their determinants are always 1 (obviously since they only rotate vectors, and do not scale them).

Matrix Diagonalization: This is mostly done in numerical computation and calculus, but I will show you how to do it later on in another chapter. The advantage of diagonalizing a matrix is the following: suppose we have a matrix $\boldsymbol{A}$ and want to take its nth power.

$$A = \begin{pmatrix} a & b \\ c & d \end{pmatrix} \rightarrow A^n = \begin{pmatrix} a & b \\ c & d \end{pmatrix}^n$$

Even if n is small this represents a large and difficult process because we would have to multiply $\boldsymbol{A}$ by itself several times under matrix multiplication. Imagine if n is bigger, if it is a hundred, it would be terrible and take a large amount of time of computation to do all this. For cases like this, it is convenient to diagonalize $\boldsymbol{A}$ into an equivalent matrix $\boldsymbol{B}$ of the form:

$$B = \begin{pmatrix} e & 0 \\ 0 & f \end{pmatrix} \rightarrow B^n = \begin{pmatrix} e & 0 \\ 0 & f \end{pmatrix}^n = \begin{pmatrix} e^n & 0 \\ 0 & f^n \end{pmatrix}$$

Which is far easier to manipulate than matrix $\boldsymbol{A}$.

Minimum Square Approximations: The reader is probably familiar with this process but it is really important when interpolating functions and simulation of many phenomena in nature. Minimum square approximation is mostly done with polynomials, however, it can be used with any function of any form.

These are the most common example of transformations. We will now discuss how transformations actually work and why.

GENERALIZING A VECTOR SPACE: A NEW LEVEL OF ABSTRACTION

We have worked with and transformed vector spaces to polynomials and matrices (thing that we could also do in the opposite way). How is this possible? At first it looks like a simple rule -- we have a transformation and apply it depending the case and needs. But this goes beyond that since there are some rules that restrain these transformations.

For example, a vector space is not equal to a polynomial, neither to a matrix. We have already defined vector spaces, but we still have to define polynomials and matrices. For this, we will take a brief trip over the horizons of abstract algebra. It's important to mention that this does not strictly belong to linear algebra, but it is good to know these things since they are the real explanation behind linear transformations.

First we will go back to arithmetic to make a concept clear. In arithmetic we learned how to add, subtract, multiply and divide, but when we got more experienced in mathematics we discovered that subtracting is actually adding a negative number and dividing is actually multiplying by the reciprocal of a number. In other words, this:

$$7-4=7+(-4) \to \textit{we are adding minus 4 here}$$

$$\frac{7}{5}=7\cdot\frac{1}{5} \to \textit{we are multiplying by the reciprocal of 5 here}$$

In other words, in arithmetic we only have two operations, addition and multiplication, of course having in mind that there exists negative numbers and reciprocals. These two terms are generalized by calling them inverses, which correspond to negative numbers under addition and reciprocals under multiplication.

A very important definition comes from the inverses, since if we have a number and add its inverse we get the identity element in addition (which is zero), and if we have a number and multiply it by its inverse we get the identity element in multiplication (which is 1). This may sound dull because we know this from arithmetic and from school. However, identity elements do not only limit themselves to usual addition and multiplication. For example, in matrices, this goes beyond that since the identities involved are the zero matrix and the identity matrix.

These concepts, with two more rules, constitute the abstract idea of a ***group***. A ***group*** is a structure with a set of elements, which can be numbers, combinations, colors or more. When studying groups it is important to use the word element because it is a more general approach, that is, more abstract. Groups come with an operation, usually denoted by an asterisk (*) which can be addition, multiplication or some other logical procedure. Groups are also closed under the operation *, which means that if we have two elements and operate them then the result is still on the group. Another fundamental requirement is that every element in the group has an inverse, and therefore the group also has an identity element. As a matter of language, to denote the operation we say "the group under the operation", for example, if the operation is addition we say "the group under addition".

To understand what a group is and learn how to identify them, let us check the sets of numbers we mentioned at the beginning of this chapter with the regular operations -- addition and multiplication:

- ***Natural numbers:*** Natural numbers are not a group under addition nor multiplication, because its elements do not have an inverse in the natural numbers. For example, the inverse of 5 under addition is -5, but -5 does not belong to the group of natural numbers (it is an integer), and the multiplicative inverse of 5 is 1/5, which does not also belong to the group of natural numbers (it is a rational number).
- ***Integers:*** Integers are a group if the operation is addition, but not if it is multiplication. If the operation is addition there will always exist an inverse of the element (for example, 5 and -5); the identity element of addition, which is 0, belongs to integers; and when you add two integers the result will always be an integer (mathematically speaking, this is called "closed under addition"), thereby making integers a group. In multiplication, however, integers are not a group because the inverse of an integer is a rational number, not an integer.

- ***Rational numbers:*** Rational numbers are a group under addition and multiplication (without the zero element) because they do satisfy the 4 conditions of a group: a set of elements which have inverses, contains the identity element and is closed under the operation (a rational number add or multiplied by another rational number gives a rational number).

It is important to take out zero from the rational numbers under multiplication because zero does not have an inverse. If zero would stay in the rational numbers under multiplication then the set would not be a group.

- ***The real numbers*** and ***complex numbers*** are a group, under addition (including zero, especially because zero is the identity element under addition) and under multiplication (excluding zero because, again, zero does not have an inverse).

Now would be a very good time to explain the ***commutative property***, which is usually omitted while explaining the properties of groups or vectors. In many cases, the commutative property applies. For example, if we have 4 apples and get two, it would be the same if have two apples and got 4 more. However, this does not always happen. For example, let us look at matrix multiplication. If we begin with a matrix A of 2 rows and 3 columns (2×3 matrix) and multiply it by a matrix B of 3 rows and 2 columns (3×2 matrix), then AB would be a 2×2 matrix. However, if would do B times A, then BA would be a 3×3 matrix. Therefore matrix multiplication, in most cases, is not commutative.

Even if A and B would be of the same size, for example, both of them would be 2×2 or $n\times n$ matrices, AB and BA would still have the same size but not equal. The only way for matrix multiplication to be equal is if both matrices have the same size and both of them are symmetrical.

Commutative property is usually shown with addition but it can also happens with multiplication. Abstract algebra generalizes this by saying that a group is Abelian if two elements operated in any order give the same result. For example, the groups mentioned above (the integers, rational, real and complex numbers) are Abelian under addition and multiplication, but matrix multiplication is not Abelian unless the two matrices are symmetrical and of the same size.

Now that we have seen this, we can check at some groups and understand how they work and why their abstraction can generalize many things. Let us look at the following examples:

- ***The integers mod 2 under addition***: Remember that every group must have at least a logical operation that satisfies the conditions explained above. The integers mod 2 are a group under addition but not under multiplication. If you are wondering, mod comes from "modular arithmetic" which is a system of arithmetic for integers. In short, the mod n operation gives the residual of dividing a number by n.

In this case, the integers mod 2 is a group built by only 2 elements, 0 and 1. This is because the residual of any number divided by 2 is either 0 or 1 as shown in the following table:

$$
\begin{array}{lll}
(0)_{mod\,2} = 0 & (1)_{mod\,2} = 1 & (2)_{mod\,2} = 0 \\
(3)_{mod\,2} = 1 & (4)_{mod\,2} = 0 & (5)_{mod\,2} = 1 \\
\vdots & \vdots & \vdots \\
(2n-2)_{mod\,2} = 0 & (2n-1)_{mod\,2} = 1 & (2n)_{mod\,2} = 0
\end{array}
$$

- As you can see we have a non-empty group, because it has two elements, 0 and 1.
- If we add two elements in the set the result is in the set. If we add 0 and 0 the result is 0, if we add 0 and 1, or 1 and 0 the result is 1 (therefore, it is an Abelian group), and if we add 1 and 1 the result is 0 (it is actually 2, but 2 mod 2 is zero).
- There exists an identity element 0.
- Each element has its inverse. The inverse of 0 is 0 and the inverse of 1 is 1.
- ***The integers mod 3 under addition***: As the same example as before, this set is a group under addition but not under multiplication. The integers mod 3 are shown in the next table:

$$(0)_{mod\,3} = 0 \qquad (1)_{mod\,3} = 1 \qquad (2)_{mod\,3} = 2$$

$$(3)_{mod\,3} = 0 \qquad (4)_{mod\,3} = 1 \qquad (5)_{mod\,3} = 2$$

$$\vdots \qquad \vdots \qquad \vdots$$

$$(3n-2)_{mod\,3} = 1 \quad (3n-1)_{mod\,3} = 2 \quad (3n)_{mod\,2} = 0$$

- As we can see we have a non-empty group, because it has three elements, 0, 1 and 2.
- If we add two elements in the set the result is in the set, if we add 0 and 0 the result is 0, if we add 0 and 1 the result is 1, 0 and 2 the result is 2, and 1 and 2 the result is 0 (this is also an Abelian group).
- There exists an identity element 0.
- Each element has its inverse. The inverse of 0 is 0, the inverse of 1 is 2 and the inverse of 2 is 1.

In general, all integers mod n are a group.

- Other groups are ***the regular polygons***, which are also called dihedral groups (dihedral comes from Ancient Greek which means “two faces”) and are one of the simplest examples of finite groups in abstract algebra. In this example, the objective is simple. Take a polygon and rotate it or flip it and assure that the polygon keeps the same shape as when it started. This may sound a little weird but let us look at the following with the simplest regular polygon, an equilateral triangle.

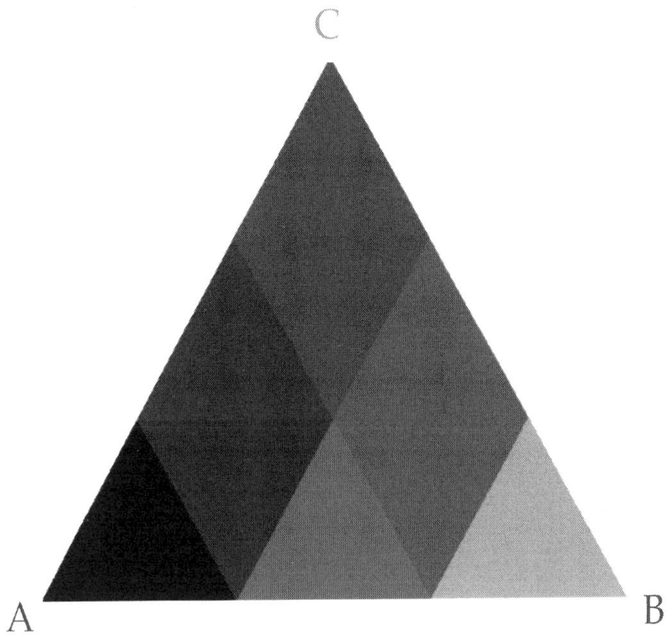

Figure 7: Equilateral original triangle.

As you can see from the above image, we have an equilateral triangle with its vertices labeled **A**, **B**, and **C**. The goal is to apply an operation that makes the triangle look the same as when it started. There are two operations that can exactly do this -- we can rotate the triangle with respect to its centroid or we can flip it with respect to its middle line. There is a third operation also but it is trivial. We will mention it at the end of this example. For now, let us focus our attention on the first operation, which is rotation. As a comment, the triangles shown in the images have a vertical line through its center. If we rotate the above triangle clockwise by 120 degrees with respect to its centroid, we get:

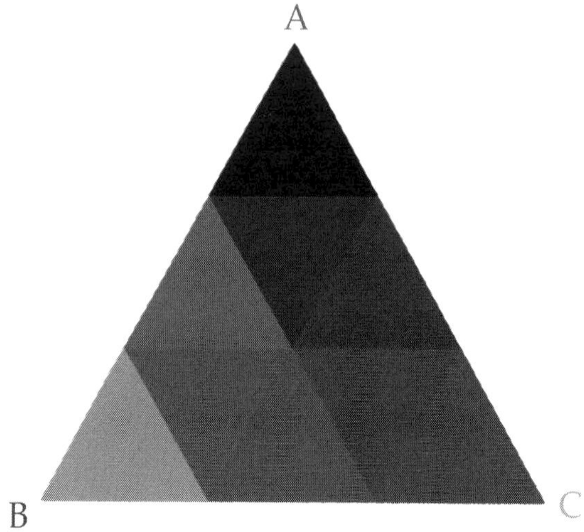

Figure 8: Triangle after one rotation.

As we can see, the triangle kept its shape. If we apply this transformation again we would get:

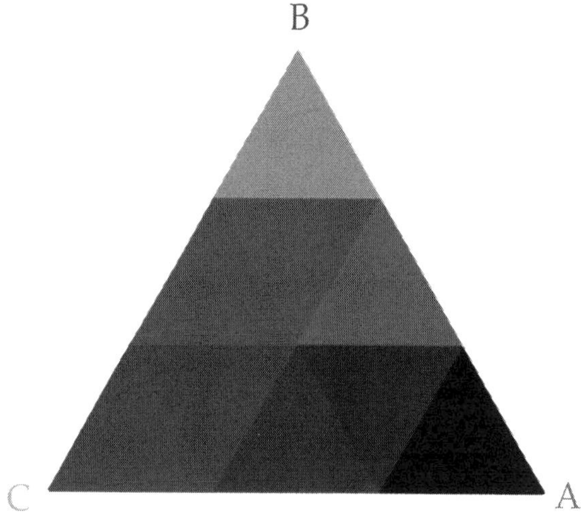

Figure 9: Triangle after two rotations.

And again, the triangle kept its shape. Let use the letter **r** to denote the operation of rotating the triangle. As we applied $\boldsymbol{r}$ twice we can denote this by $\boldsymbol{r}^2$. If we would apply the rotation again, then we would return to the triangle's original state.

Now we can proceed with the other transformation which is flipping. As the triangle is symmetrical with respect to its middle line, then flipping the triangle does not alter the shape. Let us call the flipping transformation ***f***. If we apply ***f*** to the original triangle, the result is:

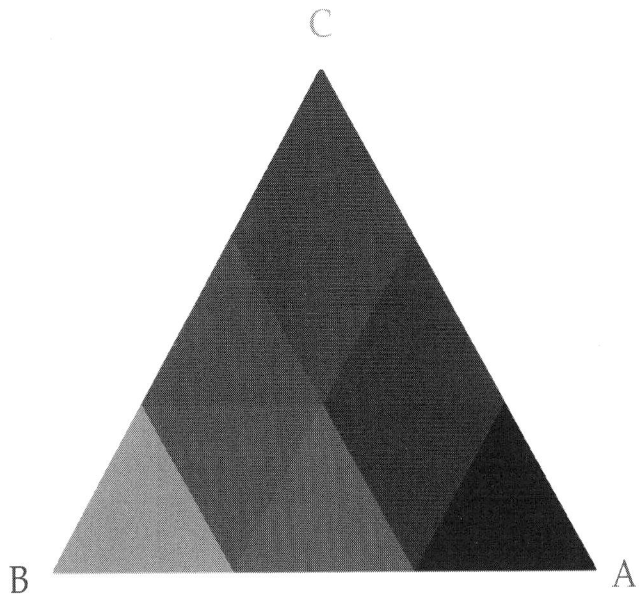

Figure 10: Triangle after a flip.

If we apply another flipping, the triangle would return to its original state. It's important to say that we could have rotated and flipped the triangle in any order, we could have applied a rotation, then a flip and then another rotation and achieve the goal since the triangle remains the same. As a fact, it's very important to note that this group is not Abelian since applying a rotation and then a flip (meaning applying ***f·r***) is not the same as applying a flip and then a rotation (meaning ***r·f***).

The final transformation that was mentioned at the beginning of this problem is the identity transformation, which does nothing to the triangle. It does not rotate it nor flip it; it leaves it the way it is. Usually this transformation in other textbooks is denoted by "e" and we will call it that way. To define the group we use the transformations shown above, transformations that change the position and orientation of vertices A, B and C. If we count A, B and C counterclockwise starting from the left vertex according to the transformation used, then we get the following:

$$ABC = e \qquad BCA = r \qquad CAB = r^2$$

$$BAC = f \qquad ACB = r \cdot f \qquad CBA = r^2 \cdot f$$

These six elements form the dihedral group for the equilateral triangle. They form a group because: (i) the group is not empty, (ii) there exists an identity element within the group, (iii) the group is closed under both operations, (iv) and each element has an inverse, which are shown next:

$$(ABC)^{-1} = e \qquad (BCA)^{-1} = r^2 \qquad (CAB)^{-1} = r$$

$$(BAC)^{-1} = f \qquad (ACB)^{-1} = r \cdot f \qquad (CBA)^{-1} = r^2 \cdot f$$

More transformations could have been done to the triangle like flipping it around any median. However, we explained two examples only just to give us an idea of how a group works or can be constructed. Other polygons represent other dihedral groups. The higher the number of sides, the bigger the group. Studying greater polygons is requires a lot of instructions, but they can show very well how to work with groups, their definitions and specially how to look for their inverses.

In abstract algebra there exists an infinite set of groups, but preceding examples have shown very well what we need to know to understand linear transformations. Going back to our topic, we said at first that we have defined vector spaces but not matrices nor polynomials. First, let us start with matrices since polynomials are a little more particular.

MATRICES

Matrices are a group, an infinite group, since it has an infinite quantity of elements. This group is so special because it is the main thing behind linear algebra. For example, let us look at matrices under addition. Matrices of $n \times m$ size form a group. Since it is not an empty set, there always exists an inverse for every element (recall that under addition the inverse of matrix A is negative A), the identity element is within the group (the zero matrix of size $n \times m$) and if we operate two matrices the result will be in the group (since adding two matrices, A and B, of size $n \times m$ will give another matrix of size $n \times m$).

Now let us look at matrices under multiplication. This group is limited since there is a condition to multiply matrices. If A is of size $n \times m$ and B is of size $m \times p$ then A·B is possible but B·A is not. Since we would have two elements which we would not be able to operate, then we restrict our group to matrices of size $n \times n$, that is, square matrices. There exists other condition, which is that all matrices in the group must have nonzero determinant. This is because a matrix with nonzero determinant has an inverse (this is also important with non-square matrices, since non-square matrices don't have an inverse).

As we have defined these two conditions we can form a group with square matrices that have non-zero determinant. They are a group because they satisfy all the conditions. This group is so important in linear algebra that it even has its own name -- the ***general linear group***. It must be noted that this group is not Abelian and that matrices do not necessarily restrain themselves to real coefficients as they can also be complex.

POLYNOMIALS

The next step is to define polynomials. **Polynomials** fall in another category than groups but very similar. Polynomials fall under the category of rings. A **ring** is another structure that has a set of elements where you may add, subtract and multiply two elements but not divide them. The conditions for a ring to exist are the same for groups except that under multiplication elements don't need an inverse.

So, to recap, a ring is a set of elements, where (i) you can add or multiply two elements and the result will stay on the ring (the ring is closed under addition and multiplication), (ii) each element has an inverse under addition but not under multiplication, and (iii) there exists the identity element. So in other words, a ring is a group that does not have an inverse under multiplication.

An example of rings is the integers -- if you add or multiply two integers (whether they are positive or negative) the result is an integer, the set of integers have 0 and 1 which are the identity elements under addition and multiplication, respectively. Non-invertible matrices also form a ring. Polynomials fall in this category since polynomials have the form

$$P(x) = a_0 + a_1x + a_2x^2 ... = \sum_{i=0}^{n} a_i x^i$$

If you add two polynomials (independently of their coefficients and degree) the result will be another polynomial. The same is true if you multiply two polynomials, but that does not necessarily happen if you divide two polynomials. Therefore, polynomials are rings.

At this point it is important to define another structure. We saw that vectors fall under vector spaces, matrices under a group, and polynomials under a ring. However, one more structure is left which is the field. A **field** is a group where addition and multiplication are commutative (in other words, Abelian). Examples of fields are the rational, real and complex numbers. Even though they fall perfectly under the definition of a group, and they are actually groups, it is better to call them fields since their operations are commutative.

All of these definitions may sound alike to each other, but their differences are important. We already know what groups, rings, and fields are, then why do vector spaces do not fall under any of these categories? Well, vectors under addition fall perfectly under the definition of a group, but vector spaces go one step further. A vector space is the union between a group of vectors under addition and a field of scalars (whether they are real or complex). This is because we can scale vectors, add them, and make linear combinations of them, etc. This union between a group and a field gives vector spaces properties beyond single groups, single rings, and single fields. This is why vector spaces are specially studied in linear algebra.

Abstract algebra is far greater than what we just saw. Actually other structures exist besides what we have discussed now, but those are beyond the scope of this book.

Going back to our discussion, we just saw how a transformation can transform a vector into a polynomial, a matrix into a vector, or a polynomial into a matrix, and vice versa. Usually at this point, many books define what is homomorphism and isomorphism.

HOMOMORPHISMS AND ISOMORPHISMS

When we apply a transformation we are going from one structure, whether it is a ring, group, vector space or anything, to another structure. To see if this two structures are alike, they must fulfill the certain conditions. Suppose our starting structure is a group called ***G*** which has an operation *, and we have a transformation ***f*** that transforms that group into another structure, let us say a ring called ***H*** with an operation ♦. The two groups are alike if homomorphism exists. Supposing we have two elements in ***G***, with names ***x*** and ***y***, then ***G*** and ***H*** are homomorphic if:

$$f(x) \blacklozenge f(y) = f(x * y)$$

Notice that the left side of the above equation is operated under ring H, and the right side is operated under group G. G and H are similar if they are **homomorphic** because both structures relate their elements with their operations. This is kind of an abstract definition, but think about it for a moment, a transformation is a homomorphism between two groups because it relates their elements in some way. Some transformations go even further with an **isomorphism,** which is actually homomorphism that is **bijective**. This means that every element in G has a single element in H and vice versa, that is, there exists an inverse transformation $\boldsymbol{f}^{(-1)}$.

LINEAR TRANSFORMATIONS: A BETTER DEFINITION

Now that we have seen all these, let us return to linear transformations. As we saw in the preceding chapter, if we have two elements, x and y, a linear transformation must fulfill the following:

$$f(x) + f(y) = f(x + y)$$

which is the definition of homomorphism when both structures have the addition operation. Therefore, every linear transformation is a homomorphism, and when applying them we can only move through similar structures, like from vector spaces to rings or groups. This is the tricky part, there are ways of expressing rings and groups as vector spaces. For example, let us start with matrices. Suppose we have an arbitrary matrix like the one shown below. We can express it like

$$\begin{pmatrix} a & b & c \\ d & e & f \end{pmatrix} = a \cdot \begin{pmatrix} 1 & 0 & 0 \\ 0 & 0 & 0 \end{pmatrix} + b \cdot \begin{pmatrix} 0 & 1 & 0 \\ 0 & 0 & 0 \end{pmatrix} + c \cdot \begin{pmatrix} 0 & 0 & 1 \\ 0 & 0 & 0 \end{pmatrix} + d \cdot \begin{pmatrix} 0 & 0 & 0 \\ 1 & 0 & 0 \end{pmatrix} + e \cdot \begin{pmatrix} 0 & 0 & 0 \\ 0 & 1 & 0 \end{pmatrix} + f \cdot \begin{pmatrix} 0 & 0 & 0 \\ 0 & 0 & 1 \end{pmatrix}$$

which is a linear combination of the basis of all 2×3 matrices, therefore the combination of a group and a field; and therefore, a vector space.

Polynomials also share this property, and they may be expressed as vector spaces. This is the power of linear algebra -- many things can be considered as vector spaces, even the solutions for homogeneous differential equations of n degree.

So, in conclusion, linear transformations take us from a starting vector space to another vector space, whether it is made by vectors, matrices, polynomials or even differential equations.

OPERATIONS WITH VECTOR SPACES

Now that we have reached here, the question arises, what can we do with vector spaces? Well, besides working with them, sometimes we would need to intersect two vector spaces, unite them or add them. Let us start with the simplest one which is addition.

Adding to vector spaces is very simple. We just need to make a linear combination of their bases. As an example, suppose we have two vector spaces ***V*** and ***W*** with the forms shown below and we want to add them

$$V = \begin{pmatrix} x \\ 0 \end{pmatrix} = x \cdot \begin{pmatrix} 1 \\ 0 \end{pmatrix}; W = \begin{pmatrix} 0 \\ y \end{pmatrix} = y \cdot \begin{pmatrix} 0 \\ 1 \end{pmatrix}$$

If we would graph these vectors in an ***xy*** plane, we would see that vectors in ***V*** are all in the ***x*** axis and vectors in ***W*** are all in the ***y*** axis. Therefore, adding these two vector spaces will enable us to represent all vectors in the ***xy*** plane. Thus,

$$V + W = x \cdot \begin{pmatrix} 1 \\ 0 \end{pmatrix} + y \cdot \begin{pmatrix} 0 \\ 1 \end{pmatrix} = \begin{pmatrix} x \\ y \end{pmatrix}$$

This is called the direct sum of two vector spaces (this term will be discussed in a short while), because we are combining the two. This is really useful and easy, but another operation like the union of two vector spaces is more restrictive. Following with the above example, if we would take the union of both vector spaces ***V*** and ***W*** the result would be all vectors in ***V*** and all vector in ***W***, that is, all vectors in the ***x*** axis and in the ***y*** axis but not a linear combination of both. In other words:

$$V \cup W = \begin{pmatrix} x \\ 0 \end{pmatrix} \wedge \begin{pmatrix} 0 \\ y \end{pmatrix} \neq \begin{pmatrix} x \\ y \end{pmatrix}$$

For example, let us look at the following expressions:

$$\vec{a} = \begin{pmatrix} 3 \\ 0 \end{pmatrix} \to \vec{a} \in V \cup W; \vec{a} \in V + W$$

$$\vec{b} = \begin{pmatrix} 0 \\ 6 \end{pmatrix} \to \vec{b} \in V \cup W; \vec{b} \in V + W$$

$$\vec{c} = \begin{pmatrix} 3 \\ 4 \end{pmatrix} \to \vec{c} \notin V \cup W; \vec{c} \in V + W$$

We may see that the union between the two vector spaces ***V*** and ***W*** are only the vectors in the ***x*** and ***y*** axis, like vectors ***a*** and ***b*** shown above, but the sum of ***V*** and ***W*** are all possible linear combinations of both vector spaces, like vector ***c*** also shown above. As a fact, the sum of two vector spaces is always a vector space but the union of two vector spaces is not necessarily another vector space. This is because when uniting two vector spaces the result will not always be closed under addition. To understand this let us look at vectors ***a***, ***b*** and ***c*** in the above statements. Vectors ***a*** and ***b*** belong in the union of vector spaces ***V*** and ***W***, but the sum of ***a*** and ***b*** does not, therefore is not closed under addition, and therefore not a vector space. However, vectors ***a***, ***b*** and ***a*** and ***b*** belong in the sum of vector spaces ***V*** and ***W***, therefore it is closed under addition, and therefore a new vector space.

Before talking about the details of the sum of vector spaces, it is important to talk about the intersection of two vector spaces first. Here is where things get interesting. We have seen before that the union of two vector spaces are all the elements of both vector spaces. Their sum is the linear combination of the elements of both vector spaces, but the intersection are all the common elements in both vector spaces. Following the preceding example with vector spaces V and W, their intersection is only one element, the zero vector. This is because the zero vector is the only common element that exists in V and W. This was a simple example, so let us look at something more creative and more general.

Suppose we have two vector spaces, denoted by the names ***U*** and ***V*** which have arbitrary bases. The basis of ***U*** will be formed by vectors $\boldsymbol{u}_1$, $\boldsymbol{u}_2$, … and $\boldsymbol{u}_i$. The basis of ***V*** will be formed by vectors $\boldsymbol{v}_1$, $\boldsymbol{v}_2$, …, $\boldsymbol{v}_j$. Therefore, the intersection of the two vector spaces will be all vectors that are common to ***U*** and ***V***. Thus

$$U \cap V = \begin{pmatrix} a_1 \\ a_2 \\ \vdots \\ a_k \end{pmatrix} \rightarrow \begin{pmatrix} a_1 \\ a_2 \\ \vdots \\ a_k \end{pmatrix} = c_1u_1 + c_2u_2 + \ldots + c_iu_i = d_1v_1 + d_2v_2 + \ldots + d_jv_j$$

An important detail with operations on vector spaces is the following. Suppose we have two arbitrary vector spaces ***U*** and ***V***; ***U*** with dimension ***n*** and ***V*** with dimension ***m*** greater than or equal to ***n***. Then the dimension of the sum of ***U*** and ***V*** will be at least equal or greater than ***m*** but equal to or less than ***n*** plus ***m***. In other words,

$$V, Uarevectorspaces.$$

$$dim(U) = n; dim(V) = m \rightarrow m \geq n$$

$$m \leq dim(U + V) \leq n + m$$

For example, if we are adding two vector spaces, ***U*** with a dimension of 2 and ***V*** with a dimension of 3, then the resulting space will have at least a dimension of 3 but less than a dimension of 5.

In case we are intersecting vector spaces then the dimension of the resulting space will be at most the smaller dimension of the summed vector spaces.

$$V, U are vector spaces.$$

$$dim(U) = n; dim(V) = m \to m \geq n$$

$$0 \leq dim(U \cap V) \leq n$$

For example, if we are intersecting two vector spaces, ***U*** with a dimension of 2 and ***V*** with a dimension of 3, then the resulting space will have at most a dimension of 2, it may be 2, 1 or 0.

Let us look at the problem below to check how this works. Suppose we have these two vector spaces and we want to intersect them

$$U = \left[\begin{pmatrix} a \\ b \\ c \end{pmatrix} \in R^3 \to a + b + c = 0 \right]$$

$$V = \left[\begin{pmatrix} x \\ y \\ z \end{pmatrix} \in R^3 \to 2x - 3y + 4z = 0 \right]$$

First we need to calculate the basis of both vector spaces, like we have done before:

$$basis V = \begin{pmatrix} x \\ y \\ z \end{pmatrix} = \begin{pmatrix} x \\ y \\ \frac{3y - 2x}{4} \end{pmatrix} = x \cdot \begin{pmatrix} 1 \\ 0 \\ -1/2 \end{pmatrix} + y \cdot \begin{pmatrix} 0 \\ 1 \\ 3/4 \end{pmatrix}$$

We now have the bases, recall that ***a, b***, ***x*** and ***y*** are constants, so we intersect both vector spaces like the following:

$$a \cdot \begin{pmatrix} 1 \\ 0 \\ -1 \end{pmatrix} + b \cdot \begin{pmatrix} 0 \\ 1 \\ -1 \end{pmatrix} = x \cdot \begin{pmatrix} 1 \\ 0 \\ -1/2 \end{pmatrix} + y \cdot \begin{pmatrix} 0 \\ 1 \\ 3/4 \end{pmatrix}$$

Which can be expressed under matrix notation like

$$\begin{pmatrix} 1 & 0 & -1 & 0 \\ 0 & 1 & 0 & -1 \\ -1 & -1 & 1/2 & -3/4 \end{pmatrix} \begin{pmatrix} a \\ b \\ x \\ y \end{pmatrix} = \begin{pmatrix} 0 \\ 0 \\ 0 \end{pmatrix}$$

Applying elemental operations we can get the matrix under echelon form

$$\begin{pmatrix} 1 & 0 & 0 & 7/2 \\ 0 & 1 & 0 & -1 \\ 0 & 0 & 1 & 7/2 \end{pmatrix} \begin{pmatrix} a \\ b \\ x \\ y \end{pmatrix} = \begin{pmatrix} 0 \\ 0 \\ 0 \end{pmatrix}$$

From which we get everything in terms of a parameter, $\boldsymbol{y}$

$$a = \frac{-7}{2} y$$

$$b = y$$

$$x = \frac{-7}{2} y$$

And substitute in the original equation

$$\frac{-7}{2} y \cdot \begin{pmatrix} 1 \\ 0 \\ -1 \end{pmatrix} + y \cdot \begin{pmatrix} 0 \\ 1 \\ -1 \end{pmatrix} = \frac{-7}{2} y \cdot \begin{pmatrix} 1 \\ 0 \\ -1/2 \end{pmatrix} + y \cdot \begin{pmatrix} 0 \\ 1 \\ 3/4 \end{pmatrix}$$

And simplify both sides of the equation to get

$$y \cdot \begin{pmatrix} -7/2 \\ 1 \\ 5/2 \end{pmatrix} = y \cdot \begin{pmatrix} -7/2 \\ 1 \\ 5/2 \end{pmatrix}$$

which is the intersection of both vector spaces. The above vector is the basis of the intersection and is the same under both sides of the equation because both vector spaces $\boldsymbol{U}$ and $\boldsymbol{V}$ can express that vector. That is exactly the definition of the intersection of vector spaces. Notice that the rule was followed, the dimension of the resulting vector space (which, in this case, is 1) is equal to or less than the smallest dimension of $\boldsymbol{U}$ and $\boldsymbol{V}$ (which, in this case, both vector spaces have a dimension of 2).

Now that we have seen the intersection between vector spaces we can discuss again the sum of two vector spaces. Suppose we have the same vector spaces ***U*** and ***V*** from the past example and suppose we want to add them. From the procedure explained previously, the resulting basis will be

$$basis U+V = a\cdot\begin{pmatrix}1\\0\\-1\end{pmatrix}+b\cdot\begin{pmatrix}0\\1\\-1\end{pmatrix}+x\cdot\begin{pmatrix}1\\0\\-1/2\end{pmatrix}+y\cdot\begin{pmatrix}0\\1\\3/4\end{pmatrix}$$

which is a new vector space of dimension 4… or not? No, actually no. Let us look at the new basis we got. It has 4 vectors in $\boldsymbol{R}^3$ which means they are a linearly dependent set of vectors, and bases must be linearly independent, therefore we have more vectors than we need. This was the detail we mentioned before when we added vector spaces -- we must check for linear dependence. The only way to add bases directly is when the intersection of the vector spaces only has the zero vector. Continuing with our problem, if we apply the Gram Schmidt orthonormalization process we get

$$\vec{v_1}=\begin{pmatrix}1\\0\\-1\end{pmatrix}\rightarrow\vec{u_1}=\frac{1}{\sqrt{2}}\begin{pmatrix}1\\0\\-1\end{pmatrix}$$

$$\vec{v_2}=\begin{pmatrix}0\\1\\-1\end{pmatrix}\rightarrow\vec{u}'_2=\vec{v_2}-\left(\vec{v_2}\cdot\vec{u_1}\right)\vec{u_1}$$

$$\vec{u}'_2=\begin{pmatrix}0\\1\\-1\end{pmatrix}-\frac{1}{2}\begin{pmatrix}1\\0\\-1\end{pmatrix}=\frac{1}{2}\begin{pmatrix}-1\\2\\-1\end{pmatrix}$$

$$\vec{u_2}=\frac{1}{\sqrt{6}}\begin{pmatrix}-1\\2\\-1\end{pmatrix}$$

$$\vec{v_3} = \begin{pmatrix} 1 \\ 0 \\ -1/2 \end{pmatrix} \rightarrow \vec{u}'_3 = \vec{v_3} - \left(\vec{v_3} \cdot \vec{u_1}\right)\vec{u_1} - \left(\vec{v_3} \cdot \vec{u_2}\right)\vec{u_2}$$

$$\vec{u}'_3 = \begin{pmatrix} 1 \\ 0 \\ -1/2 \end{pmatrix} - \frac{3}{4}\begin{pmatrix} 1 \\ 0 \\ -1 \end{pmatrix} - \frac{(-1)}{12}\begin{pmatrix} -1 \\ 2 \\ -1 \end{pmatrix} = \frac{1}{6}\begin{pmatrix} 1 \\ 1 \\ 1 \end{pmatrix}$$

$$\vec{u_3} = \frac{1}{\sqrt{3}}\begin{pmatrix} 1 \\ 1 \\ 1 \end{pmatrix}$$

$$\vec{v_4} = \begin{pmatrix} 0 \\ 1 \\ 3/4 \end{pmatrix} \rightarrow \vec{u}'_4 = \vec{v_4} - \left(\vec{v_4} \cdot \vec{u_1}\right)\vec{u_1} - \left(\vec{v_4} \cdot \vec{u_2}\right)\vec{u_2} - \left(\vec{v_4} \cdot \vec{u_3}\right)\vec{u_3}$$

$$\vec{u}'_4 = \begin{pmatrix} 0 \\ 1 \\ 3/4 \end{pmatrix} - \frac{(-3)}{8}\begin{pmatrix} 1 \\ 0 \\ -1 \end{pmatrix} - \frac{5}{24}\begin{pmatrix} -1 \\ 2 \\ -1 \end{pmatrix} - \frac{7}{12}\begin{pmatrix} 1 \\ 1 \\ 1 \end{pmatrix} = \begin{pmatrix} 0 \\ 0 \\ 0 \end{pmatrix}$$

$$\vec{u_4} = \vec{0}$$

Therefore, we only have 3 linearly independent vectors and our sum of vector spaces has a dimension of 3, which is higher than the highest dimension of the original vector spaces ***U*** and ***V*** (which is 2) but less than the sum of the dimensions of the original vector spaces ***U*** and ***V*** (which is 2+2=4).

By the way, when working two vector spaces the following property must be fulfilled, having two vector spaces ***U*** and ***V*** then:

$$dim(U+V) + dim(U \cap V) = dim(U) + dim(V)$$

Which is read as "the dimension of the addition of ***U*** and ***V*** plus the dimension of the intersection of ***U*** and ***V*** is equal to the dimension of

U plus the dimension of ***V***." You can check this property with the past example.

As an interesting fact, the term echelon formation comes from the french word *échelon* which means "rung of a ladder." It is also an arrangement of troops in parallel lines, like we see in echelon form matrices.

CHAPTER 3

MATRICES

"Concern should drive us into action and not into a depression. No man is free who cannot control himself"

Pythagoras

INTRODUCTION

Matrices are the center field of study in Linear Algebra. They are very useful in many senses, and serve well to express most of the problems concerning humanity. Besides matrices have been going around the earth for the past two thousand years, first recorded in China in the book "Nine Chapters of the Mathematical Art", where there is even a section with a very similar process to Gaussian elimination that served to solve systems of equations before Seki and Leibniz worked with them around the second half of the seventeenth century. However, their knowledge was not preserved until then, when they were formally presented to the mathematical community.

It is of relevance to mention that most of the advances have been happening in the past decades with matrix notation and its use, along with several definitions and discoveries because of a common effort between millions of mathematicians, economists, physicists, programmers and engineers around the globe. This is because matrix notation is very easy, useful and intuitive to use – the very same reasons why it has become standard to use matrix notation for problems like systems of equations even when they are not linear. Matrices present a nice and clean way to represent problems and have many more advantages, which we will discuss in this chapter.

MATRICES

We saw in the past chapter that matrices can be considered as vector spaces. However, this is just one of their many definitions. It is true that defining matrices as a vector space is the most abstract definition that generalizes their properties because, as vectors, matrices are abstract entities that shape themselves to the needs of the user and his insight, but abstraction elevates our thoughts to levels where sometimes we do not touch the ground, or even reality. Therefore, let us talk about matrices from their first steps.

The first thing we are introduced with while learning about matrix notation is that they are extremely, useful for solving systems of linear equations since they are an arrangement of data. Matrix notations are an arrangement that separates three main things: **coefficients**, **variables** and **independent terms**. Even when the system is small, like a two by two system, it gives a clean look on what is going on in the system. And as variables and sizes get bigger, matrices remain clean and allow to express whole systems of equations as a single equation in the following way:

$$A\vec{x} = \vec{b}$$

Let us look at this with higher detail in the next example.

$$a_1x + a_2y = b_1$$
$$a_3x + a_4y = b_2$$

wherein we have two equations with two variables, **x** and **y**. To get a cleaner view of the system, it is good to apply matrix notation. Take note that the matrix notation being used now is a convention used. For example, in the Chinese tablet from two thousand years ago variables were not written in the arrangement and the independent terms were put below the coefficients just like this

$a_1 \quad a_3$

$a_2 \quad a_4$

$b_1 \quad b_2$

which is a valid form of arrangement. It is probably not the best since we don't see the variables anywhere, but it is valid. However, the usual notation used nowadays is highly developed and more refined since it has its own rules (which we will discuss later) and logic. But the fundamental key in which matrix notation and all of its operations resides is that the variables are listed vertically at the right of the coefficient matrix, starting with the first variable at the top and finishing with the last variable at the bottom. You should already be familiar with matrix notation by now, but the importance of mentioning this will be noticed further ahead. Applying matrix notation gets us:

$$\begin{pmatrix} a_1 & a_2 \\ a_3 & a_4 \end{pmatrix} \begin{pmatrix} x \\ y \end{pmatrix} = \begin{pmatrix} b_1 \\ b_2 \end{pmatrix}$$

This use for a matrix is the most basic one, the most simple of all, but many things can be done from it. We can see we have two equations that relate x and y in some form. Two equations which are separate, and each of them gives us valuable **information** on the system. The detail that they are separate is key since we can work with one independently of the other, of course, while maintaining the equality. For example, we could take one of the equations and multiply it by a scalar on both sides, that is, multiply one of the rows of the matrix by a scalar. Doing so would not change the information the equation gives us, it would just transform it into an equivalent form. Having two separate equations also gives us an advantage that we would not have if we only have a single equation. But since we can combine both equations, we can also combine the information they give us. This can be done by adding them or subtracting them to get a new equation like the following:

$$\begin{pmatrix} a_1 & a_2 \\ a_3 & a_4 \\ a_1 + a_3 & a_2 + a_4 \end{pmatrix} \begin{pmatrix} x \\ y \end{pmatrix} = \begin{pmatrix} b_1 \\ b_2 \\ b_1 + b_2 \end{pmatrix}$$

However, adding and subtracting a given set of equations like we just did does not help us at all since it does not gives us any other information from what we had. We must be clever when working with matrices, since problems of high complexity are well simplified this way, but they also may get worse if matrices are used the wrong way. With a great matrix comes great genius on how to use it.

Thus, in solving a system of equations,

- we can ***multiply*** rows by a certain scalar maintaining the equality;
- we can ***add*** or ***subtract*** rows to get more useful data; and
- although it was not mentioned before, we can also ***switch*** rows from one to another.

Any of these operations does not alter the system, it just takes it into an equivalent form. These are the so called elemental operations.

You should already be very familiar on how to solve systems of equations. We just mentioned these comments because they are needed for the next steps. As such, we can express the solution for the generic system of equations above:

$$\begin{pmatrix} x \\ y \end{pmatrix} = \frac{1}{a_1 \cdot a_4 - a_2 \cdot a_3} \begin{pmatrix} a_4 & -a_2 \\ -a_3 & a_1 \end{pmatrix} \begin{pmatrix} b_1 \\ b_2 \end{pmatrix}$$

ROWS AND COLUMNS

Matrices are built by two main things -- rows and columns, and each of them has very interesting properties. Let us look at the following matrix multiplied by a vector in the following system of equations:

$$\begin{pmatrix} a_1 & b_1 & c_1 \\ a_2 & b_2 & c_2 \\ a_3 & b_3 & c_3 \end{pmatrix} \begin{pmatrix} x \\ y \\ z \end{pmatrix} = \begin{pmatrix} b_1 \\ b_2 \\ b_3 \end{pmatrix}$$

which can be rewritten as

$$x \cdot \begin{pmatrix} a_1 \\ a_2 \\ a_3 \end{pmatrix} + y \cdot \begin{pmatrix} b_1 \\ b_2 \\ b_3 \end{pmatrix} + z \cdot \begin{pmatrix} c_1 \\ c_2 \\ c_3 \end{pmatrix} = \begin{pmatrix} b_1 \\ b_2 \\ b_3 \end{pmatrix}$$

We can see that the system of equations above is actually a linear combination of the columns of the coefficient matrix. It is strange to think about it that way, because logically we are only able to do elemental operations on rows and not on columns. In other words, we can multiply and add rows but we cannot do the same with columns. However, the fact that the independent term is equal to a linear combination of the columns of the coefficient matrix is necessary.

Matrices, as we said, have rows and columns. Each row is called a row vector and each column is called a column vector. The set of row vectors and column vectors represent a certain space, which are called the row space and the column space, respectively. The row space are the rows of the coefficient matrix, which give us the information we need to solve systems of equations since each row is an equation that relates the variables that we are after. The column space are the columns of the coefficient matrix, which give us information on the linear independence of the column vectors of the matrix and their result.

Each space, the row space and the column space, has its own properties that we will see further ahead, but one of them is this. The independent term of the system of equations is always a linear combination of the column space, therefore the solution depends on how the column vectors are combined to get that independent term.

It is true that when working with systems of equations we can only do elemental operations on rows, which changes column space, too (because we are changing the vectors). But as we change the vectors we should also be changing the independent term in the same way (maintaining the equality). Something like this:

$$\begin{pmatrix} 1 & 2 \\ 3 & 4 \end{pmatrix} \begin{pmatrix} x \\ y \end{pmatrix} = \begin{pmatrix} 3 \\ 7 \end{pmatrix}$$

$$\begin{pmatrix}1\\3\end{pmatrix};\begin{pmatrix}2\\4\end{pmatrix} \rightarrow \textit{vectors of the column space}$$

$$\begin{pmatrix}3\\7\end{pmatrix} \rightarrow \textit{independent term of the system of equations}$$

We see that we have two vectors, and that a certain linear combination of them will give us the independent term of the system we are working on. To get the value of the variables that makes a linear combination of the vectors equal to the independent term of the of the system, we need to reduce the matrix into its echelon form by applying elemental operations. For example, if we subtract from the second row 3 times the first one, we get:

$$\begin{pmatrix}1 & 2\\0 & -2\end{pmatrix}\begin{pmatrix}x\\y\end{pmatrix}=\begin{pmatrix}3\\-2\end{pmatrix}$$

From what we can see, every vector has changed. The independent term has changed in the same way also, thereby maintaining the equality. As a final comment on this example, the solution is $\boldsymbol{x}=\boldsymbol{y}=1$.

GENERAL SYSTEMS OF EQUATIONS

Usually, examples are given with square matrices, since they are one of the most desirable cases. But let us discuss the more general case, rectangular systems, like the following of size

$$\begin{pmatrix}a_{11} & a_{12} & \cdots & a_{1m}\\ a_{21} & a_{22} & \cdots & a_{2m}\\ \vdots & \vdots & \cdots & \vdots\\ a_{n1} & a_{n2} & \cdots & a_{nm}\end{pmatrix}\begin{pmatrix}x_1\\x_2\\ \vdots\\x_m\end{pmatrix}=\begin{pmatrix}b_1\\b_2\\ \vdots\\b_n\end{pmatrix}$$

As you can see, we have a coefficient matrix (let us call it $\boldsymbol{A}$) of size . Note that the number of rows of the matrix is equal to the number of equations (we have $\boldsymbol{n}$ equations), and that the number of columns is equal to the number of variables (we have $\boldsymbol{m}$ variables). Let us play with $\boldsymbol{m}$ and $\boldsymbol{n}$.

If n is greater than m, that is, we have more equations than variables, that would mean that our system might not have a solution since the information the equations are giving could contradict itself, like in the following example:

$$\begin{pmatrix} 1 & 2 \\ 3 & 4 \\ 5 & 6 \end{pmatrix} \begin{pmatrix} x \\ y \end{pmatrix} = \begin{pmatrix} 10 \\ 11 \\ 9 \end{pmatrix} \rightarrow No\,solution$$

This is the most undesirable case, and the most complicated one.

If m is greater than n, that is, we have more variables than equations, that would mean that our system would have infinite solutions since the equations are not giving enough information to get a unique solution and we would have something that is called a free variable just like this

$$\begin{pmatrix} 1 & 2 & 3 \\ 4 & 5 & 6 \end{pmatrix} \begin{pmatrix} x \\ y \\ z \end{pmatrix} = \begin{pmatrix} 10 \\ 11 \end{pmatrix} \rightarrow Infinite\,solutions$$

If m is equal to n, then we could only have one solution to the system like some of the examples we have seen so far. However, this conclusion does not depend only on the size of the matrix; it depends on n and m. It also depends on another definition which is the ***rank*** of the matrix.

RANK OF A MATRIX

The rank of a matrix is a concept similar to the dimension of a vector space since it is the number of linearly independent rows in a system of equations and we could see it as the dimension of the row space. For example, let us use one of the examples that we had before.

$$\begin{pmatrix} a_1 & a_2 \\ a_3 & a_4 \\ a_1 + a_3 & a_2 + a_4 \end{pmatrix} \begin{pmatrix} x \\ y \end{pmatrix} = \begin{pmatrix} b_1 \\ b_2 \\ b_1 + b_2 \end{pmatrix}$$

We can see in the set that the third equation is exactly the sum of the first equation and the second equation. Thus, the set of equations is linearly dependent and one equation does not give us any new information. In case we apply row operations to the third row in the above system, we could subtract from it the first and second row to get a row full of zeros, therefore this kinds of rows can be taken out of the system, like this:

$$\begin{pmatrix} a_1 & a_2 \\ a_3 & a_4 \\ 0 & 0 \end{pmatrix} \begin{pmatrix} x \\ y \end{pmatrix} = \begin{pmatrix} b_1 \\ b_2 \\ 0 \end{pmatrix}$$

$$\begin{pmatrix} a_1 & a_2 \\ a_3 & a_4 \end{pmatrix} \begin{pmatrix} x \\ y \end{pmatrix} = \begin{pmatrix} b_1 \\ b_2 \end{pmatrix}$$

It is important to see that a row of zeros in the row space represent extra equations that are useless. However, sometimes they represent a danger to the system itself, since a row of zeros should always equal zero. If a row of zeros does not equal zero, then the system would contradict itself and would not have a solution.

The usual way to check the linearly independent rows of a matrix, that is, the rank of a matrix, is to reduce it by elemental operations to echelon form. The number rows different from zero is the rank of the matrix. Let us look at the following example to see that:

$$\begin{pmatrix} 1 & 2 & 3 \\ 4 & 6 & 8 \\ 3 & 4 & 5 \end{pmatrix} \begin{pmatrix} x \\ y \\ z \end{pmatrix} = \begin{pmatrix} 5 \\ 12 \\ 7 \end{pmatrix}$$

If we reduce the matrix to its echelon form, we get:

$$\begin{pmatrix} 1 & 0 & -1 \\ 0 & 1 & 2 \\ 0 & 0 & 0 \end{pmatrix} \begin{pmatrix} x \\ y \\ z \end{pmatrix} = \begin{pmatrix} -3 \\ 4 \\ 0 \end{pmatrix}$$

Where we can see that we have a row of zeros and two linearly independent equations, and therefore, the rank is 2. Thus, we actually have a 2×3 system of equations, which means that we have an infinite number of possible combinations of the column space to get the independent term of the system. We have infinite combinations because we have a ***free*** variable. Even though we have an infinite number of solutions they will have a pattern that is correlated with the ***null space*** of the coefficient matrix.

THE NULL SPACE

We have said that matrices are a way of transforming vectors, and when transforming vectors we are going from one vector space to another vector space. We also said that the zero element must be contained within the spaces we are traveling through, and that the transformation of the zero element in the input space must always be the zero element of the output space.

But we have also said that some transformations are able to transform two different and independent vectors into the same one. But what vectors can we transform such that their output is the zero vector? The answer to that question is the null space, and we will see why the null space of a transformation is very important.

Suppose we have a system of equations shown in the following way:

$$A\vec{x} = \vec{b}$$

A regular system of equations, which can also be seen as a matrix ***A*** which transforms a vector ***x*** into a vector ***b***. As we are introducing the concept of the null space, let us remember that the null space is the set of vectors whose transformation is the zero vector. Therefore, we may express the above equation in the following way.

If ***v*** is a vector that belongs in the null space, then it is true that:

$$A(\vec{x} + \vec{v}) = \vec{b} + \vec{0} = \vec{b}$$

Therefore, the null space is always part of the solution of any system of equations. Now that we have said this, we will show how to calculate the null space of a transformation, that is, of a matrix. Let us continue with the example we had.

If we want all vectors which transformed become the zero vector, then:

$$\begin{pmatrix} 1 & 0 & -1 \\ 0 & 1 & 2 \end{pmatrix} \begin{pmatrix} x \\ y \\ z \end{pmatrix} = \begin{pmatrix} 0 \\ 0 \end{pmatrix}$$

From here we can see that:

$$y = -2z$$
$$x = z$$

Therefore, all vectors of the form shown below will have the zero vector for transformation:

$$\vec{v} = z\begin{pmatrix} 1 \\ -2 \\ 1 \end{pmatrix} \to T(\vec{v}) = \vec{0}$$

The null space is very important because it gives us all the possible solutions to systems of equations like as explained above, because we will have a particular solution plus any solution that belongs to the null space. The particular solution appears because of the independent term in the system of equations.

This can be better shown with the full preceding example:

$$\begin{pmatrix} 1 & 0 & -1 \\ 0 & 1 & 2 \end{pmatrix}\begin{pmatrix} x \\ y \\ z \end{pmatrix} = \begin{pmatrix} -3 \\ 4 \end{pmatrix}$$

This can be rewritten as:

$$x\cdot\begin{pmatrix} 1 \\ 0 \end{pmatrix} + y\cdot\begin{pmatrix} 0 \\ 1 \end{pmatrix} + z\cdot\begin{pmatrix} -1 \\ 2 \end{pmatrix} = \begin{pmatrix} -3 \\ 4 \end{pmatrix}$$

Where we can see that the column vectors of the coefficient matrix are linearly dependent on each other. However, in this case, we cannot remove a column since it would removing a whole variable of the system. Therefore, we will do the following arrangement:

$$x\cdot\begin{pmatrix} 1 \\ 0 \end{pmatrix} + y\cdot\begin{pmatrix} 0 \\ 1 \end{pmatrix} = \begin{pmatrix} -3 \\ 4 \end{pmatrix} + z\cdot\begin{pmatrix} 1 \\ -2 \end{pmatrix}$$

From where we get:

$$x = -3 + z$$
$$y = 4 - 2z$$

And the final solution will be:

$$solution = \begin{pmatrix} x \\ y \\ z \end{pmatrix} = \begin{pmatrix} -3+z \\ 4-2z \\ z \end{pmatrix} = \begin{pmatrix} -3 \\ 4 \\ 0 \end{pmatrix} + z \begin{pmatrix} 1 \\ -2 \\ 1 \end{pmatrix} = solution_{particular} + null\ space$$

The null space is always a part of the solution for a system of equations. However, in systems wherein the number of linearly independent equations equals the number of variables, the null space is only the zero vector, and therefore, does not appear.

Here is another information about the null space. The null space is a subspace, and every subspace has a dimension. The dimension of the null space is equal to the number of free variables in the system, since those free variables are able to change freely to annul the other ones in the system, therefore making the zero vector.

Another important detail is that the dimension of the null space is called the **nullity**, and that the number of free variables in the system is equal to the total number of variables minus the number of linearly independent rows. The number of independent rows is equal to the **rank** of the matrix. From here comes a very important theorem in linear algebra, which is the ***rank nullity*** theorem. Which says that:

$$A = A_{n \times m} \rightarrow Rank(A) + Null(A) = m$$

Which is read as, "if ***A*** is a matrix of size then the rank of ***A*** plus the nullity of ***A*** is equal to ***m*,** being ***m*** the number of columns in the matrix." The rank of a matrix is not only equal to the number of ***linearly independent rows***, but also to the number of ***linearly independent columns***. We will show the proof of this below.

Suppose we have a generic matrix of size with the form shown below:

$$\begin{pmatrix} a_{11} & a_{12} & \cdots & a_{1m} \\ a_{21} & a_{22} & \cdots & a_{2m} \\ \vdots & \vdots & \cdots & \vdots \\ a_{n1} & a_{n2} & \cdots & a_{nm} \end{pmatrix}$$

And by applying elemental operations we changed it into an equivalent matrix of the following form:

$$\begin{pmatrix} 1 & 0 & 0 & \cdots & 0 & b_{1,r+1} & \cdots & b_{1,m} \\ 0 & 1 & 0 & \cdots & 0 & b_{2,r+1} & \cdots & b_{2,m} \\ \vdots & \vdots & \vdots & \cdots & \vdots & \vdots & \cdots & \vdots \\ 0 & 0 & 0 & \cdots & 1 & b_{r,r+1} & \cdots & b_{r,m} \\ 0 & 0 & 0 & \cdots & 0 & 0 & \cdots & 0 \\ 0 & 0 & 0 & \cdots & 0 & 0 & \cdots & 0 \end{pmatrix}$$

Where we have ***r*** rows in echelon form. The ***b*** elements on the right are the coefficients that cannot be eliminated by applying elemental operations. If we take the zero ***n-r*** rows of zeros from the bottom we will get

$$\begin{pmatrix} 1 & 0 & 0 & \cdots & 0 & b_{1,r+1} & \cdots & b_{1,m} \\ 0 & 1 & 0 & \cdots & 0 & b_{2,r+1} & \cdots & b_{2,m} \\ \vdots & \vdots & \vdots & \cdots & \vdots & \vdots & \cdots & \vdots \\ 0 & 0 & 0 & \cdots & 1 & b_{r,r+1} & \cdots & b_{r,m} \end{pmatrix}$$

But as you can see we have ***r*** columns of linearly independent vectors, and ***m-r*** other column vectors that can be rewritten as a linear combination of the first ***r*** ones (this last ***m-r*** columns also represent the number of free variables on the system). Therefore, from this approach, we can see that the rank of the matrix is not only equal to the number of linearly independent rows but is also equal to the number of linearly independent column in the matrix. For this, the rank is also called the **column rank**. In conclusion,

If ***A*** is a matrix of size with rank ***r***, then:

- The rank of the matrix is equal to the number of linearly independent rows and columns in the matrix.
- If ***A*** is part of a system of equations, the number of free variables is equal to ***m-r*** which is also equal to the nullity of A.
- If all rows or all columns of A are linearly independent, that is, if the rank is equal to ***n*** or ***m***, then the matrix is called full rank.
- If the rank is equal to the number of variables on a system of equations, then there only exists a ***unique*** solution.
- If the rank is less than the number of variables on a system of equations, then there are ***infinite*** solutions.

The rank will never be higher than the number of variables in the system, but in case we have more rows than variables in the system, that could correspond with the no solution situation on a system of equations. This, again, is dangerous because in case we have a row of zeros which does not equate to zero, then the system of equations does not have a solution because it contradicts itself as shown in the example below.

$$\begin{pmatrix} 1 & 2 \\ 3 & 4 \\ 5 & 6 \end{pmatrix} \begin{pmatrix} x \\ y \end{pmatrix} = \begin{pmatrix} 10 \\ 11 \\ 9 \end{pmatrix}$$

Applying elemental operations, we get

$$\begin{pmatrix} 1 & 0 \\ 0 & 1 \\ 0 & 0 \end{pmatrix} \begin{pmatrix} x \\ y \end{pmatrix} = \begin{pmatrix} -9 \\ 19/2 \\ -3 \end{pmatrix}$$

Now that we have seen this, let us recap some things about the ***column space*** and the ***row space***:

- The ***row space*** gives us information on how the variables are correlated and the possibility of a solution.
- The ***column space*** gives us information about the null space, and that the independent term of the system of equations is a linear combination of the column vectors.
- Both the ***column*** and ***row*** spaces have the same dimension, which is the ***rank*** of the matrix and depending on the ***rank***, the ***number of rows*** and the ***number of columns*** we might know what kind of solutions we could get from the system.

If you understand these concepts very well, then it would not be difficult to solve the following problem:

Suppose we have an unknown matrix ***A***, and we have the following conditions:

$$A\vec{x} = \begin{pmatrix} 1 \\ 0 \\ 0 \end{pmatrix} \rightarrow \textit{has no solution};\ A\vec{x} = \begin{pmatrix} 0 \\ 1 \\ 0 \end{pmatrix} \rightarrow \textit{has a single solution}$$

The question involves getting as much information we can get from matrix ***A***, which includes its number of rows, its number of columns, and its rank.

Just by looking at it, we can see that the number of rows of ***A*** is 3 since it is equal to the number of components of the independent term. From the first condition (the no solution condition) we see that the first row of ***A*** should be a row of zeros (because a row of zeros that equates to a nonzero number means that a solution does not exist), and therefore the rank should be at most 2. The third condition gives information on the columns of ***A***, since a single solution means that the number of variables is equal to the rank, and the number of variables is equal to the number of columns of ***A***, Therefore, the columns of ***A*** are at most 2.

With this information we could conclude that ***A*** could have the following forms:

$$A = \begin{pmatrix} 0 & 0 \\ 1 & 0 \\ 0 & 1 \end{pmatrix} \to Rank = columns = 2 A = \begin{pmatrix} 0 \\ 1 \\ 0 \end{pmatrix} \to Rank = columns = 1$$

where both examples satisfy all the conditions given above.

PROPERTIES OF MATRICES

Working with matrices is quite intuitive, because we may add, subtract and multiply matrices. To demonstrate how addition works in matrices is very easy, but to demonstrate how matrix multiplication works is more difficult.

The proof below shows how matrix multiplication works. Suppose we have vectors ***u***, ***v*** and ***w***, and matrices ***A*** and ***B*** arranged in the following way:

$$\vec{w} = B\vec{v}$$

$$\vec{v} = A\vec{u}$$

$$\vec{w} = BA\vec{u}$$

We wish to know the form of a matrix ***C***, such that:

$$\vec{w} = C\vec{u}$$

$$C = BA$$

From other courses in algebra we already know how matrix multiplication is done, but we will give a proof of it, which will be very helpful to demonstrate other properties. Let us suppose matrices ***A*** and ***B***, and vectors ***u***, ***v*** and ***w***, have the following size and coefficients.

$$A = A_{n\times m} = \begin{pmatrix} A_{11} & A_{12} & \cdots & A_{1m} \\ A_{21} & A_{22} & \cdots & A_{2m} \\ \vdots & \vdots & \cdots & \vdots \\ A_{n1} & A_{n2} & \cdots & A_{nm} \end{pmatrix} \quad \vec{v} = v_{n\times 1} = \begin{pmatrix} v_1 \\ v_2 \\ \vdots \\ v_n \end{pmatrix} = A\vec{u}$$

$$\vec{u} = u_{m\times 1} = \begin{pmatrix} u_1 \\ u_2 \\ \vdots \\ u_m \end{pmatrix}$$

$$B = B_{p\times n} = \begin{pmatrix} B_{11} & B_{12} & \cdots & B_{1n} \\ B_{21} & B_{22} & \cdots & B_{2n} \\ \vdots & \vdots & \cdots & \vdots \\ B_{p1} & B_{p2} & \cdots & B_{pn} \end{pmatrix}$$

$$\vec{w} = w_{p\times 1} = \begin{pmatrix} v_1 \\ v_2 \\ \vdots \\ v_p \end{pmatrix}$$

Let us start by looking at how the coefficients of vector ***v*** will be. From the convention used in matrix notation we know that the ***i-th*** component of ***v*** will be the ***i-th*** row of ***A*** multiplied by vector ***u***:

$$v_i = \sum_{j=1}^{j=m} A_{ij} u_j \rightarrow 1 \le i \le n$$

Since we have an equation for the elements in vector ***v***, now we can find an expression for the ***k-th*** element in vector ***w***, which will be the ***k-th*** row of B multiplied by vector v.

$$w_k = \sum_{i=1}^{i=n} B_{ki} v_i = \sum_{i=1}^{i=n} B_{ki} \left(\sum_{j=1}^{j=m} A_{ij} u_j \right) \rightarrow \begin{matrix} 1 \le i \le n \\ 1 \le k \le p \end{matrix}$$

If we expand the sums and rearrange,

$$w_k = \sum_{i=1}^{i=n} B_{ki}\left(\sum_{j=1}^{j=m} A_{ij}u_j\right) = \begin{matrix} B_{k1}\cdot\left(A_{11}\cdot u_1 + A_{12}\cdot u_2 + \ldots + A_{1m}\cdot u_m\right) \\ +B_{k2}\cdot\left(A_{21}\cdot u_1 + A_{22}\cdot u_2 + \ldots + A_{2m}\cdot u_m\right) \\ \vdots \\ +B_{kn}\cdot\left(A_{n1}\cdot u_1 + A_{n2}\cdot u_2 + \ldots + A_{nm}\cdot u_m\right) \end{matrix}$$

$$w_k = \sum_{i=1}^{i=n} B_{ki}\left(\sum_{j=1}^{j=m} A_{ij}u_j\right) = \begin{matrix} u_1\cdot\left(B_{k1}\cdot A_{11} + B_{k2}\cdot A_{21} + \ldots + B_{kn}\cdot A_{n1}\right) \\ +u_2\cdot\left(B_{k1}\cdot A_{12} + B_{k2}\cdot A_{22} + \ldots + B_{kn}\cdot A_{n2}\right) \\ \vdots \\ +u_m\cdot\left(B_{k1}\cdot A_{1m} + B_{k2}\cdot A_{2m} + \ldots + B_{kn}\cdot A_{nm}\right) \end{matrix}$$

Therefore,

$$w_k = \sum_{j=1}^{j=m}\left(\sum_{i=1}^{i=n} B_{ki}\cdot A_{ij}\right)u_j \to 1 \le k \le pw_k = \sum_{j=1}^{j=m} C_{kj}u_j \to 1 \le k \le p$$

Therefore,

$$C_{kj} = \sum_{i=1}^{i=n} B_{ki}\cdot A_{ij} \to \begin{matrix} 1 \le k \le p \\ 1 \le j \le m \end{matrix}$$

And there we have it, the proof for matrix multiplication, which is really useful to demonstrate other properties, specially the inverse matrix and the transpose matrix.

TRANSPOSE MATRIX AND INVERSE MATRIX

The transpose of a matrix is an operator that flips all the elements in the matrix over the main diagonal of the matrix. An example is this:

$$A = \begin{pmatrix} 1 & 2 & 3 \\ 4 & 5 & 6 \end{pmatrix} \to A^T = \begin{pmatrix} 1 & 4 \\ 2 & 5 \\ 3 & 6 \end{pmatrix}$$

Therefore if we transpose a matrix of size we would get a new matrix of size . Since a transpose is an operator, therefore, it has its own properties. And from what we have seen, we can prove all of them. The properties of the transpose are:

Given two matrices ***A*** and ***B***, and a scalar ***c***, then they must fulfill the following:

$$\left(A^T\right)^T = A$$

$$\left(A+B\right)^T = A^T + B^T$$

$$\left(cA\right)^T = c\cdot A^T$$

$$\left(A\cdot B\right)^T = B^T\cdot A^T$$

Obviously ***A*** and ***B*** must be of proper size for these properties to work. Under addition ***A*** and ***B*** must have the same size, and under multiplication the number of columns of ***A*** must be equal to the number of rows of ***B.***

However, explaining ***the inverse of a matrix*** is more complex. Remember that the inverse of a matrix is another matrix such that when they are multiplied the result is the identity matrix. To show how to calculate the inverse of a matrix, let us keep with the notation we had for matrix multiplication:

$$C_{kj} = \sum_{i=1}^{i=n} B_{ki}\cdot A_{ij} \rightarrow \begin{matrix} 1 \le k \le p \\ 1 \le j \le m \end{matrix}$$

For example, let us suppose ***A*** is our initial matrix and ***B*** the inverse matrix of ***A***. If we treat ***C*** to be the identity matrix we get the following system of equations so we can find the inverse of ***A***. It's important to say that for this to happen ***p*** and ***m*** must be of the same size since the identity matrix is a square matrix:

$$\begin{matrix} C_{11} = B_{11}\cdot A_{11} + B_{12}\cdot A_{21} + \ldots + B_{1n}\cdot A_{n1} = 1 \\ C_{12} = B_{11}\cdot A_{12} + B_{12}\cdot A_{22} + \ldots + B_{1n}\cdot A_{n2} = 0 \\ \vdots \\ C_{1m} = B_{11}\cdot A_{1m} + B_{12}\cdot A_{2m} + \ldots + B_{1n}\cdot A_{nm} = 0 \end{matrix} \rightarrow \begin{pmatrix} A_{11} & A_{21} & \cdots & A_{n1} \\ A_{12} & A_{22} & \cdots & A_{n2} \\ \vdots & \vdots & \vdots & \vdots \\ A_{1m} & A_{2m} & \cdots & A_{nm} \end{pmatrix}\begin{pmatrix} B_{11} \\ B_{12} \\ \vdots \\ B_{1n} \end{pmatrix} = \begin{pmatrix} 1 \\ 0 \\ \vdots \\ 0 \end{pmatrix}$$

$$\begin{aligned} C_{21} &= B_{21} \cdot A_{11} + B_{22} \cdot A_{21} + \ldots + B_{2n} \cdot A_{n1} = 0 \\ C_{22} &= B_{21} \cdot A_{12} + B_{22} \cdot A_{22} + \ldots + B_{2n} \cdot A_{n2} = 1 \\ &\vdots \\ C_{2m} &= B_{21} \cdot A_{1m} + B_{22} \cdot A_{2m} + \ldots + B_{2n} \cdot A_{nm} = 0 \end{aligned} \rightarrow \begin{pmatrix} A_{11} & A_{21} & \cdots & A_{n1} \\ A_{12} & A_{22} & \cdots & A_{n2} \\ \vdots & \vdots & \vdots & \vdots \\ A_{1m} & A_{2m} & \cdots & A_{nm} \end{pmatrix} \begin{pmatrix} B_{21} \\ B_{22} \\ \vdots \\ B_{2n} \end{pmatrix} = \begin{pmatrix} 0 \\ 1 \\ \vdots \\ 0 \end{pmatrix}$$

$$\vdots$$

$$\begin{aligned} C_{m1} &= B_{m1} \cdot A_{11} + B_{m2} \cdot A_{21} + \ldots + B_{mn} \cdot A_{n1} = 0 \\ C_{m2} &= B_{m1} \cdot A_{12} + B_{m2} \cdot A_{22} + \ldots + B_{mn} \cdot A_{n2} = 0 \\ &\vdots \\ C_{mm} &= B_{m1} \cdot A_{1m} + B_{m2} \cdot A_{2m} + \ldots + B_{mn} \cdot A_{nm} = 1 \end{aligned} \rightarrow \begin{pmatrix} A_{11} & A_{21} & \cdots & A_{n1} \\ A_{12} & A_{22} & \cdots & A_{n2} \\ \vdots & \vdots & \vdots & \vdots \\ A_{1m} & A_{2m} & \cdots & A_{nm} \end{pmatrix} \begin{pmatrix} B_{m1} \\ B_{m2} \\ \vdots \\ B_{mn} \end{pmatrix} = \begin{pmatrix} 0 \\ 0 \\ \vdots \\ 1 \end{pmatrix}$$

Thus, we have ***m*** systems of equations, each with ***m*** equations and ***n*** variables, where each system of equation represents the coefficients of a row of matrix ***B***. Let us suppose ***A*** and ***B*** are ***full rank*** (if not, then they would not have an inverse matrix). Let us discuss the possibilities that would arise depending on the values of ***m*** and ***n***.

- ***if m is greater than n,*** then we know we might not have a solution because we would have more equations than variables.
- ***if m is equal to n,*** then there only exists a single solution; and
- ***if n is greater than m,*** then they are infinite solutions.

Let us stay for now with the case in which ***m*** is equal to ***n***, so ***A*** and ***B*** are both square matrices.

Solving those enormous systems of equations could be a titanic adventure, but as we mentioned in the beginning we must be clever using matrices. We may have two approaches to solve for all coefficients in ***B***, solve the ***m*** system of equations separately, or we could solve all of them at the same time (since every system has the matrix coefficients). We mentioned that every system represents the coefficients of a row of matrix ***B***. However, ***Carl Friedrich Gauss*** had an amazing approach.

GAUSS INVERSE METHOD

Gauss saw that a way to solve the systems of linear equations was by applying elemental operations to find the coefficients of the inverse matrix we are after, in a way we will discuss now.

Elemental operations are row operations that we have already discussed, so in order to find the inverse we want, we need to look for a way to operate the rows of the coefficient matrix to get it into its echelon form. For this we use a simple fact about matrices. We saw that ***multiplying a matrix by a vector*** gives a linear combination of the columns, but ***multiplying a vector by a matrix*** gives a linear combination of the rows of the matrix. We can see this in the following:

$$A = A_{n\times m} = \begin{pmatrix} A_{11} & A_{12} & \cdots & A_{1m} \\ A_{21} & A_{22} & \cdots & A_{2m} \\ \vdots & \vdots & \cdots & \vdots \\ A_{n1} & A_{n2} & \cdots & A_{nm} \end{pmatrix}$$

$$\vec{u} = u_{m\times 1} = \begin{pmatrix} u_1 \\ u_2 \\ \vdots \\ u_m \end{pmatrix} \qquad \vec{v} = v_{1\times n} = \begin{pmatrix} v_1 & v_2 & \dots & v_n \end{pmatrix}$$

$$A\vec{u} = u_1 \cdot \begin{pmatrix} A_{11} \\ A_{21} \\ \vdots \\ A_{n1} \end{pmatrix} + u_2 \cdot \begin{pmatrix} A_{12} \\ A_{22} \\ \vdots \\ A_{n2} \end{pmatrix} + \dots + u_m \cdot \begin{pmatrix} A_{1m} \\ A_{2m} \\ \vdots \\ A_{nm} \end{pmatrix} \qquad \vec{v}A = \begin{matrix} v_1 \cdot \begin{pmatrix} A_{11} & A_{12} & \cdots & A_{1m} \end{pmatrix} \\ +v_2 \cdot \begin{pmatrix} A_{21} & A_{22} & \cdots & A_{2m} \end{pmatrix} \\ \vdots \\ +v_n \cdot \begin{pmatrix} A_{n1} & A_{n2} & \cdots & A_{nm} \end{pmatrix} \end{matrix}$$

This example was just with a vector of 1 row, but let us take this to a higher level. We must keep focusing on our problem, which is to find a matrix ***B*** such that it is the inverse of our matrix ***A***. We already discussed that ***A*** and ***B*** are full rank and that they are square matrices of size. This is shown in the following problem:

$$\begin{pmatrix} B_{11} & B_{12} & \cdots & B_{1n} \\ B_{21} & B_{22} & \cdots & B_{2n} \\ \vdots & \vdots & \cdots & \vdots \\ B_{n1} & B_{n2} & \cdots & B_{nn} \end{pmatrix} \cdot \begin{pmatrix} A_{11} & A_{12} & \cdots & A_{1n} \\ A_{21} & A_{22} & \cdots & A_{2n} \\ \vdots & \vdots & \cdots & \vdots \\ A_{n1} & A_{n2} & \cdots & A_{nn} \end{pmatrix} = \begin{pmatrix} 1 & 0 & \cdots & 0 \\ 0 & 1 & \cdots & 0 \\ \vdots & \vdots & \cdots & \vdots \\ 0 & 0 & \cdots & 1 \end{pmatrix}$$

But having in mind the idea we had before of a linear combination of the rows, we can see matrix ***B*** as a way of combining the rows of ***A*** such that we get the identity, where the ith row of ***B*** would represent the way in which the rows of ***A*** are linearly combined to get the ith row of the identity matrix. The same question remains, which is to find the coefficients of matrix **B** but in a different way, since we now know that they represent how the rows of **A** will be linearly combined to get the identity matrix. This linear combinations are the row operations we need to apply to matrix **A** to obtain the identity matrix, that is, reduce **A** to its echelon form.

The question is how to compute and record all the operations needed to reduce matrix **A** into its echelon form. The answer to that question is ***Gauss's method*** of finding the inverse.

Gauss's method of finding the inverse is simple. If you write matrix **A** at one side and the identity matrix at the other side, just apply row operations to reduce matrix **A** into its echelon form and do the same row operations to the identity matrix. When the process is finished, the identity matrix will be where matrix **A** originally was, and what is left where the identity matrix originally was, is now the inverse of **A**.

This method corresponds to what we just saw on linearly combining the rows of **A** such that we reduce it to its echelon form. Since by applying the same row operations that we do on matrix **A** to the identity, the identity matrix records this by becoming a new matrix. It is a little difficult to understand this, so let us look at it with equations.

Remember our matrix **A** has the following form and coefficients:

$$A = \begin{pmatrix} A_{11} & A_{12} & \cdots & A_{1n} \\ A_{21} & A_{22} & \cdots & A_{2n} \\ \vdots & \vdots & \cdots & \vdots \\ A_{n1} & A_{n2} & \cdots & A_{nn} \end{pmatrix}$$

By applying Gauss's method, we write matrix **A** at one side (let us do it on the left) and the identity matrix on the other side (therefore, on the right), like this:

$$\left(\begin{matrix} A_{11} & A_{12} & \cdots & A_{1n} \\ A_{21} & A_{22} & \cdots & A_{2n} \\ \vdots & \vdots & \cdots & \vdots \\ A_{n1} & A_{n2} & \cdots & A_{nn} \end{matrix}\right]\left[\begin{matrix} 1 & 0 & \cdots & 0 \\ 0 & 1 & \cdots & 0 \\ \vdots & \vdots & \cdots & \vdots \\ 0 & 0 & \cdots & 1 \end{matrix}\right)$$

We now apply the same row operations to both matrices. Let us suppose we divide the first row by , getting the following:

$$\left(\begin{array}{cccc|cccc} 1 & A_{12}/A_{11} & \cdots & A_{1n}/A_{11} & 1/A_{11} & 0 & \cdots & 0 \\ A_{21} & A_{22} & \cdots & A_{2n} & 0 & 1 & \cdots & 0 \\ \vdots & \vdots & \cdots & \vdots & \vdots & \vdots & \cdots & \vdots \\ A_{n1} & A_{n2} & \cdots & A_{nn} & 0 & 0 & \cdots & 1 \end{array}\right)$$

We can see that both matrices have changed, but the identity matrix on the right has recorded our operation. Therefore our problem is now solved, because we found a way to record all the operations we need to do to matrix $\boldsymbol{A}$ so that we reduce it to its echelon form. To check this, look at the following. If we multiply the matrix we have on the ***right*** by our original matrix $\boldsymbol{A}$, then we will get the matrix we have on the ***left***:

$$\begin{pmatrix} 1/A_{11} & 0 & \cdots & 0 \\ 0 & 1 & \cdots & 0 \\ \vdots & \vdots & \cdots & \vdots \\ 0 & 0 & \cdots & 1 \end{pmatrix} \cdot \begin{pmatrix} A_{11} & A_{12} & \cdots & A_{1n} \\ A_{21} & A_{22} & \cdots & A_{2n} \\ \vdots & \vdots & \cdots & \vdots \\ A_{n1} & A_{n2} & \cdots & A_{nn} \end{pmatrix} = \begin{pmatrix} 1 & A_{12}/A_{11} & \cdots & A_{1n}/A_{11} \\ A_{21} & A_{22} & \cdots & A_{2n} \\ \vdots & \vdots & \cdots & \vdots \\ A_{n1} & A_{n2} & \cdots & A_{nn} \end{pmatrix}$$

This property will hold for any operation we do, since the identity matrix will record every elemental operation we make on matrix A. Therefore when we apply enough row operations to reduce A into its echelon form, then what is left on the right side is the inverse of A.

This was Gauss approach to find inverses, which was immensely clever and far easier than Laplace's approach. However, Laplace was a genius for discovering his method.

To conclude this section let us make an example:

Suppose we have the following matrix $\boldsymbol{A}$ and we want to calculate its inverse.

$$A = \begin{pmatrix} 10 & 12 & 7 & 4 & 9 \\ 5 & 6 & 4 & 2 & 3 \\ 7 & 8 & 1 & 0 & 5 \\ 0 & 17 & 6 & 0 & 2 \\ 0 & 11 & 13 & 9 & 7 \end{pmatrix}$$

Applying *Gauss's method*,

$$\left(\begin{array}{ccccc|ccccc} 10 & 12 & 7 & 4 & 9 & 1 & 0 & 0 & 0 & 0 \\ 5 & 6 & 4 & 2 & 3 & 0 & 1 & 0 & 0 & 0 \\ 7 & 8 & 1 & 0 & 5 & 0 & 0 & 1 & 0 & 0 \\ 0 & 17 & 6 & 0 & 2 & 0 & 0 & 0 & 1 & 0 \\ 0 & 11 & 13 & 9 & 7 & 0 & 0 & 0 & 0 & 1 \end{array}\right) \to \begin{array}{c} R_1 = R_1/10 \\ R_2 = R_2 - 5R_1 \\ R_3 = R_3 - 7R_1 \end{array}$$

$$\left(\begin{array}{ccccc|ccccc} 1 & 6/5 & 7/10 & 2/5 & 9/10 & 1/10 & 0 & 0 & 0 & 0 \\ 0 & 0 & 1/2 & 0 & -3/2 & -1/2 & 1 & 0 & 0 & 0 \\ 0 & -2/5 & -39/10 & -14/5 & -13/10 & -7/10 & 0 & 1 & 0 & 0 \\ 0 & 17 & 6 & 0 & 2 & 0 & 0 & 0 & 1 & 0 \\ 0 & 11 & 13 & 9 & 7 & 0 & 0 & 0 & 0 & 1 \end{array}\right)$$

We see that we do not have a pivot in the second column, therefore, we can exchange two rows of the matrix and continue reducing it:

$$\left(\begin{array}{ccccc|ccccc} 1 & 6/5 & 7/10 & 2/5 & 9/10 & 1/10 & 0 & 0 & 0 & 0 \\ 0 & 0 & 1/2 & 0 & -3/2 & -1/2 & 1 & 0 & 0 & 0 \\ 0 & -2/5 & -39/10 & -14/5 & -13/10 & -7/10 & 0 & 1 & 0 & 0 \\ 0 & 17 & 6 & 0 & 2 & 0 & 0 & 0 & 1 & 0 \\ 0 & 11 & 13 & 9 & 7 & 0 & 0 & 0 & 0 & 1 \end{array}\right) \to \begin{array}{c} R_2 \Leftrightarrow R_3 \\ R_2 = \dfrac{-5}{2} R_2 \\ R_4 = R_4 - 17R_2 \\ R_5 = R_5 - 11R_2 \end{array}$$

$$\left(\begin{array}{ccccc|ccccc} 1 & 6/5 & 7/10 & 2/5 & 9/10 & 1/10 & 0 & 0 & 0 & 0 \\ 0 & 1 & 39/4 & 7 & 13/4 & 7/4 & 0 & -5/2 & 0 & 0 \\ 0 & 0 & 1/2 & 0 & -3/2 & -1/2 & 1 & 0 & 0 & 0 \\ 0 & 0 & -639/4 & -119 & -213/4 & -119/4 & 0 & 85/2 & 1 & 0 \\ 0 & 0 & -377/4 & -68 & -115/4 & -77/4 & 0 & 55/2 & 0 & 1 \end{array}\right) \to \begin{array}{c} R_3 = 2R_3 \\ R_4 = R_4 + \dfrac{639}{4} R_3 \\ R_5 = R_5 + \dfrac{377}{4} R_3 \end{array}$$

We can see coefficients are getting difficult to manage, however, let us continue quickly. Following the last operations we get:

$$\left(\begin{array}{ccccc|ccccc} 1 & 6/5 & 7/10 & 2/5 & 9/10 & 1/10 & 0 & 0 & 0 & 0 \\ 0 & 1 & 39/4 & 7 & 13/4 & 7/4 & 0 & -5/2 & 0 & 0 \\ 0 & 0 & 1 & 0 & -3 & -1 & 2 & 0 & 0 & 0 \\ 0 & 0 & 0 & -119 & -1065/2 & -379/2 & 639/2 & 85/2 & 1 & 0 \\ 0 & 0 & 0 & -68 & -623/2 & -227/2 & 377/2 & 55/2 & 0 & 1 \end{array}\right) \to \begin{array}{c} R_4 = R_4 - R_5 \\ R_4 = \dfrac{-1}{51} R_4 \\ R_5 = R_5 + 68R_4 \end{array}$$

$$\left(\begin{array}{ccccc|ccccc} 1 & 6/5 & 7/10 & 2/5 & 9/10 & 1/10 & 0 & 0 & 0 & 0 \\ 0 & 1 & 39/4 & 7 & 13/4 & 7/4 & 0 & -5/2 & 0 & 0 \\ 0 & 0 & 1 & 0 & -3 & -1 & 2 & 0 & 0 & 0 \\ 0 & 0 & 0 & 1 & 13/3 & 76/51 & -131/51 & -5/17 & -1/51 & 1/51 \\ 0 & 0 & 0 & 0 & -101/6 & -73/6 & 83/6 & 15/2 & -4/3 & 7/3 \end{array}\right) \to R_5 = \frac{-6}{101} R_5$$

$$\left(\begin{matrix} 1 & 6/5 & 7/10 & 2/5 & 9/10 \\ 0 & 1 & 39/4 & 7 & 13/4 \\ 0 & 0 & 1 & 0 & -3 \\ 0 & 0 & 0 & 1 & 13/3 \\ 0 & 0 & 0 & 0 & 1 \end{matrix}\right|\left|\begin{matrix} 1/10 & 0 & 0 & 0 & 0 \\ 7/4 & 0 & -5/2 & 0 & 0 \\ -1 & 2 & 0 & 0 & 0 \\ 76/51 & -131/51 & -5/17 & -1/51 & 1/51 \\ 73/101 & -83/101 & -45/101 & 8/101 & -14/101 \end{matrix}\right)$$

And from here we could do row reduction upwards to have ***the identity on the left side and the inverse of A on the right side***. However, one of the reasons all of these procedures were done is to show what we said before, if we multiply the matrix on the right by matrix ***A*****,** we will get the matrix on the left.

Let us say the matrices we have on the right are called matrices ***R***, and the matrices we have on the left are called matrices ***L***, so:

Some examples of these are some that we just did:

$$\begin{pmatrix} 1/10 & 0 & 0 & 0 & 0 \\ -1/2 & 1 & 0 & 0 & 0 \\ -7/10 & 0 & 1 & 0 & 0 \\ 0 & 0 & 0 & 1 & 0 \\ 0 & 0 & 0 & 0 & 1 \end{pmatrix} \cdot \begin{pmatrix} 10 & 12 & 7 & 4 & 9 \\ 5 & 6 & 4 & 2 & 3 \\ 7 & 8 & 1 & 0 & 5 \\ 0 & 17 & 6 & 0 & 2 \\ 0 & 11 & 13 & 9 & 7 \end{pmatrix} = \begin{pmatrix} 1 & 6/5 & 7/10 & 2/5 & 9/10 \\ 0 & 0 & 1/2 & 0 & -3/2 \\ 0 & -2/5 & -39/10 & -14/5 & -13/10 \\ 0 & 17 & 6 & 0 & 2 \\ 0 & 11 & 13 & 9 & 7 \end{pmatrix}$$

$$\begin{pmatrix} 1/10 & 0 & 0 & 0 & 0 \\ 7/4 & 0 & -5/2 & 0 & 0 \\ -1/2 & 1 & 0 & 0 & 0 \\ -119/4 & 0 & 85/2 & 1 & 0 \\ -77/4 & 0 & 55/2 & 0 & 1 \end{pmatrix} \begin{pmatrix} 10 & 12 & 7 & 4 & 9 \\ 5 & 6 & 4 & 2 & 3 \\ 7 & 8 & 1 & 0 & 5 \\ 0 & 17 & 6 & 0 & 2 \\ 0 & 11 & 13 & 9 & 7 \end{pmatrix} = \begin{pmatrix} 1 & 6/5 & 7/10 & 2/5 & 9/10 \\ 0 & 1 & 39/4 & 7 & 13/4 \\ 0 & 0 & 1/2 & 0 & -3/2 \\ 0 & 0 & -639/4 & -119 & -213/4 \\ 0 & 0 & -377/4 & -68 & -115/4 \end{pmatrix}$$

$$\begin{pmatrix} 1/10 & 0 & 0 & 0 & 0 \\ 7/4 & 0 & -5/2 & 0 & 0 \\ -1 & 2 & 0 & 0 & 0 \\ 76/51 & -131/51 & -5/17 & -1/51 & 1/51 \\ 73/101 & -83/101 & -45/101 & 8/101 & -14/101 \end{pmatrix} \cdot \begin{pmatrix} 10 & 12 & 7 & 4 & 9 \\ 5 & 6 & 4 & 2 & 3 \\ 7 & 8 & 1 & 0 & 5 \\ 0 & 17 & 6 & 0 & 2 \\ 0 & 11 & 13 & 9 & 7 \end{pmatrix} = \begin{pmatrix} 1 & 6/5 & 7/10 & 2/5 & 9/10 \\ 0 & 1 & 39/4 & 7 & 13/4 \\ 0 & 0 & 1 & 0 & -3 \\ 0 & 0 & 0 & 1 & 13/3 \\ 0 & 0 & 0 & 0 & 1 \end{pmatrix}$$

To finish this problem we would continue reducing the matrix. However, we will only show the result:

$$A^{-1} = \frac{1}{1717}\begin{pmatrix} -197 & 610 & 91 & -88 & -48 \\ -854 & 448 & 900 & -59 & 280 \\ 2006 & -799 & -2295 & 408 & -714 \\ -2819 & 1704 & 2810 & -623 & 1065 \\ 1241 & -1411 & -765 & 136 & -238 \end{pmatrix}$$

OTHER MATRIX INVERSES

We just saw the case where ***square matrices*** have inverses, obviously assuming that they are full rank, because if they are not full rank then there are linearly dependent columns which means the matrix does not have a one-to-one correspondence after transforming a vector, and therefore, does not have an inverse and as consequence its determinant is zero. There are like 8 equivalent reasons for a square matrix not to have an inverse. However, we will now focus on matrices that are not square, that is, ***rectangular matrices***.

In early algebra courses, we are usually told that rectangular matrices do not have inverses, but that is not actually true. Rectangular matrices do have inverses, but only certain type of inverses. Let's take a look at the following example:

$$A = \begin{pmatrix} 1 & 0 \\ 0 & 1 \\ 0 & 1 \end{pmatrix}; A^{-1} = \begin{pmatrix} 1 & 0 & 0 \\ 0 & 1/2 & 1/2 \end{pmatrix}; A^{-1} \cdot A = \begin{pmatrix} 1 & 0 \\ 0 & 1 \end{pmatrix}$$

From what we can see, rectangular matrices do have inverses. But as we said, they only have a certain type of inverses. For this let us take this problem analytically. Suppose we have an $n \times m$ matrix called $\boldsymbol{A}$ of the following way:

$$A = A_{n\times m} = \begin{pmatrix} A_{11} & A_{12} & \cdots & A_{1m} \\ A_{21} & A_{22} & \cdots & A_{2m} \\ \vdots & \vdots & \cdots & \vdots \\ A_{n1} & A_{n2} & \cdots & A_{nm} \end{pmatrix}$$

The idea is to compute another matrix ***B*** of size $m \times n$ such that when they are multiplied the identity matrix shows up. It is important that the sizes of matrices ***A*** and ***B*** are $n \times m$ and $m \times n$, respectively, because when we multiply them, the result will be square which is what identity matrices are. Let us assume that both matrices ***A*** and ***B*** are full rank.

$$A = A_{n\times m}; B = B_{m\times n}$$

$$A \cdot B = A \cdot B_{n\times n}$$

$$B \cdot A = B \cdot A_{m\times m}$$

Let us remember our goal, we want to find matrices ***A*** and ***B*** such that when they are multiplied, the result is the identity matrix. The key condition here is that the columns and rows of the identity matrix are linearly independent (we will show why this is a key factor during the explanation).

- If $m > n$ then $A \cdot B$ will be the identity matrix of siz $n \times n$ e, but $B \cdot A$ will not be the identity matrix of size $m \times m$. This is because ***A*** has ***n*** linearly independent rows, and ***B*** has ***n*** linearly independent columns, therefore, $A \cdot B$ can be seen as the linearly independent rows of ***A*** transforming the linearly independent columns of ***B.*** Thus, the result is also a set of linearly independent vectors. In conclusion, with the right choice of ***B***, $A \cdot B$ will be the identity matrix.

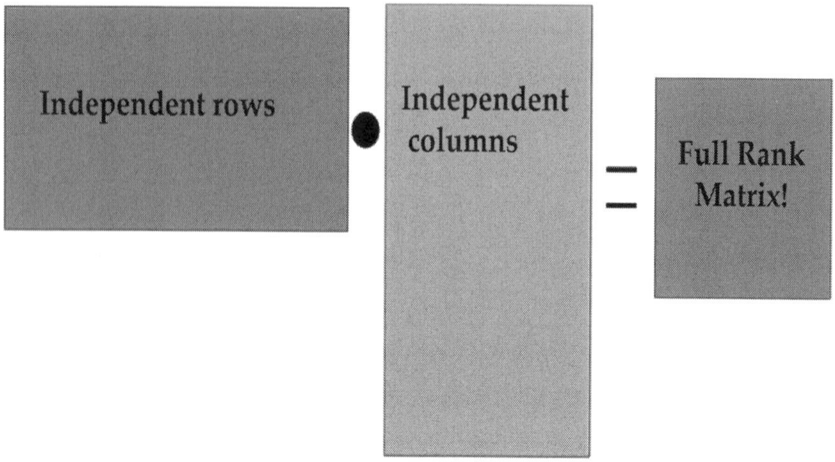

Figure 11: Full rank matrix.

However, $B \cdot A$ will not be able to do this, because ***B*** has linearly dependent rows and ***A*** has linearly dependent columns. When the linearly dependent rows of ***B*** transform the linearly dependent columns of ***A*** the result will be linearly dependent. The new matrix will not be the identity because the columns of the identity are linearly independent.

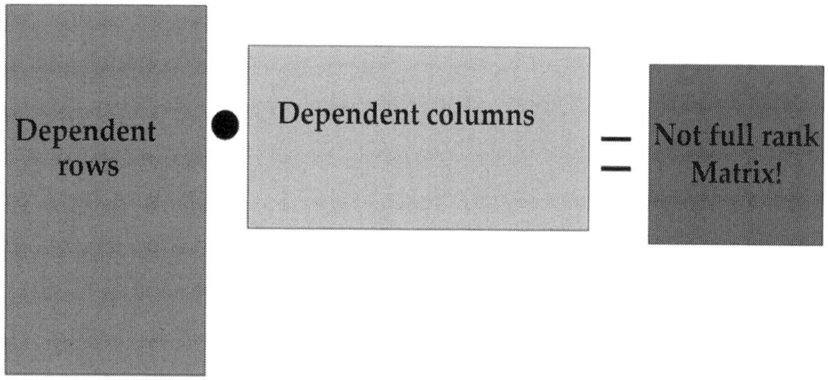

Figure 12: Not a full rank matrix.

The opposite happens if $m < n$. So, to recap, if ***A*** is a $n \times m$ matrix, and ***B*** is a matrix of $m \times n$ such that it is the inverse of ***A***, then:

$$if\ m > n \rightarrow A \cdot B = I_{n \times n}; B \cdot A \neq I_{m \times m}$$

$$if\ m < n \rightarrow B \cdot A = I_{m \times m}; A \cdot B \neq I_{n \times n}$$

That was the easy part. Now, we need to compute a way to obtain ***B*** in terms of ***A***. We saw this problem when both matrices were square. However, we now want to know how this is when they are not square matrices. We may remember that having rectangular matrices then computing an inverse gave ***more variables than equations***. And when solving this kind of systems of equations we saw that if we have more variables than equations then we would be able to find ***infinite solutions***. However, we want a nice way to compute the inverse of matrix ***A***, or even better, we want the best solution to the inverse of matrix ***A***.

The inverse of ***A*** must depend only on ***A*** and nothing else. For this, mathematicians made the following arrangement:

$$A = A_{n \times m}$$

$$if\ m > n$$

$$A \cdot A^T = A \cdot A^T{}_{n \times n}\ will\,be\,invertible$$

$$\left(A \cdot A^T\right) \cdot \left(A \cdot A^T\right)^{-1} = I_{n \times n} = A \cdot A^{-1}$$

$$A^{-1} = A^T \left(A \cdot A^T\right)^{-1}$$

This is called the ***right inverse***, because if $m > n$ then we can only multiply the inverse on the right. To get an expression for an inverse when $m < n$, the same procedure is applied:

$$A = A_{n \times m}$$

$$if m < n$$

$$A^T \cdot A = A^T \cdot A_{m \times m} will be invertible$$

$$\left(A^T \cdot A\right)^{-1} \cdot \left(A^T \cdot A\right) = I_{m \times m} = A^{-1} \cdot A$$

$$A^{-1} = \left(A^T \cdot A\right)^{-1} \cdot A^T$$

This is called the ***left inverse*** because when $m < n$, then we can only multiply the inverse on the left.

Besides the reason exposed before, multiplying $\boldsymbol{A}$ by its transpose has several benefits because the result is guaranteed to be a ***symmetrical matrix***, which is a matrix easier to work with and its rank will be equal to the rank of matrix ***A***. So to recap:

$$A = A_{n \times m}$$

$$if m > n \rightarrow A^{-1} = A^T \cdot \left(A \cdot A^T\right)^{-1}$$

$$if n > m \rightarrow A^{-1} = \left(A^T \cdot A\right)^{-1} \cdot A^T$$

In the first case where $m > n$, A^{-1} is a ***right inverse***, and in the second case $m < n$, A^{-1} is a ***left inverse***. This way we can have inverses of rectangular matrices, again when they are full rank.

CHAPTER 4

WORKING WITH MATRICES

"My first feeling was that there was no way to continue. Writing isn't like math; in math, two plus two always equals four no matter what your mood is like. With writing, the way you feel changes everything."

Stephenie Meyer

INTRODUCTION

Theory tells us how things will work, but that does not mean people do that in real life. Theory and practice always get along, since practice cannot go against theory. However, nature has a hidden and unknown way to practice theory at its finest, to produce the most beautiful and common phenomena we see in real life. Unfortunately, humans still do not find a way to do it the same way as nature.

For this we must require methods, solutions and arrangements that deal with this -- deal with errors, deal with the impossibility of calculations, deal with the difficulties of calculating a value that is impossible to calculate at first sight. In this chapter we will see real and simple methods that do this. However, there are more sophisticated and improved methods that exist today.

Nevertheless, each tool must be used according to the problem. If we have to kill a fly, we would not do it with a rifle, and if we would have to kill a beast we would definitely not kill it with a shoe. For this reason, this methods are still used in situations of low level requirements.

LEAST SQUARES

We have talked and explained with high detail systems with equal rank to number of variables and rank less than the number of variables. However, the rank of a system could never be higher than the number of variables. Then what happens when we have more linearly independent equations than variables?

We have discussed that we might not have a solution in this case, but another approach, from what we saw in past chapters, if the rows of the coefficient matrix are greater than its columns we could apply its left inverse and solve the problem, always remembering that matrix ***A*** has no free variables.

This can be done even if we get a row of zeros equal to a nonzero digit when we reduce a matrix ***A*** to its echelon form. Let us find out why using the following example:

$$\begin{pmatrix} 1 & 2 \\ 3 & 4 \\ 5 & 6 \end{pmatrix} \begin{pmatrix} x \\ y \end{pmatrix} = \begin{pmatrix} 10 \\ 11 \\ 9 \end{pmatrix}$$

We saw that this problem does not have a solution. However, what happens if we apply the left inverse?

$$\begin{pmatrix} 1 & 3 & 5 \\ 2 & 4 & 6 \end{pmatrix} \cdot \begin{pmatrix} 1 & 2 \\ 3 & 4 \\ 5 & 6 \end{pmatrix} \begin{pmatrix} x \\ y \end{pmatrix} = \begin{pmatrix} 1 & 3 & 5 \\ 2 & 4 & 6 \end{pmatrix} \cdot \begin{pmatrix} 10 \\ 11 \\ 9 \end{pmatrix}$$

$$\begin{pmatrix} 35 & 44 \\ 44 & 56 \end{pmatrix} \begin{pmatrix} x \\ y \end{pmatrix} = \begin{pmatrix} 88 \\ 118 \end{pmatrix}$$

$$\begin{pmatrix} x \\ y \end{pmatrix} = \frac{1}{35 \cdot 56 - 44^2} \cdot \begin{pmatrix} 56 & -44 \\ -44 & 35 \end{pmatrix} \cdot \begin{pmatrix} 88 \\ 118 \end{pmatrix}$$

$$\begin{pmatrix} x \\ y \end{pmatrix} = \frac{1}{24} \cdot \begin{pmatrix} -264 \\ 258 \end{pmatrix} = \begin{pmatrix} -11 \\ 10,75 \end{pmatrix}$$

If we apply the left inverse we get a solution. How is that possible? Let us introduce the values we got into the original equations:

$$\begin{pmatrix} 1 & 2 \\ 3 & 4 \\ 5 & 6 \end{pmatrix} \begin{pmatrix} -11 \\ 10,75 \end{pmatrix} = \begin{pmatrix} 10,5 \\ 10 \\ 9,5 \end{pmatrix} \neq \begin{pmatrix} 10 \\ 11 \\ 9 \end{pmatrix}$$

We can see that the solutions we got are not exactly equal to the independent term, but they are very close to it. This is because the system really does not have a solution. It is impossible to get a solution that exactly satisfies all of the equations, even though we managed to get a solution by multiplying by the inverse of the coefficient matrix.

What we got from here is an approximation to the solution. We know we don't have a solution, but in case we could have one, we would want to get the best one. This is what this procedure does and we will show why.

As we have said before, the independent term of the system of equations should always be a linear combination of the column vectors of the coefficient matrix. This means that the independent term should be in the column space. In the above example, we can see this more easily since the original system of equations has 2 column vectors (that is, the column space is a plane). Our system does not have a solution since the independent term does not belong to the column space. Graphically it would look like this:

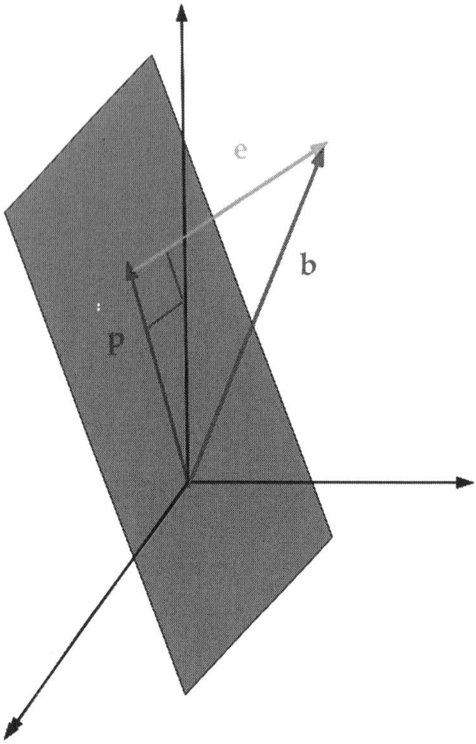

Figure 13: Least Squares approximation.

We can see a blue plane that represents our column space, and a red vector ***b*** that represents our independent term. As you can see, in this case, vector ***b*** does not belong in the column space. This means that we don't have a solution. However, in case we would want the best solution we could get, we would have to find the projection of ***b*** over the column space, which in the above graph is vector ***p*** (colored in blue). Obviously, projecting ***b*** over the column space will generate an error which is vector ***e*** (colored in green). All of these conditions (the best solution condition and the projection) imply that the error vector ***e*** must be perpendicular to the column space. Because vector ***e*** is minimum when ***b*** and ***e*** are perpendicular, and it is the difference between the independent term we have (vector ***b***) minus the independent term we can get (vector ***p***), therefore,:

$$\vec{e} = \vec{b} - \vec{p}$$

So to recap, we have this system of equations that does not have a solution:

$$A\vec{x} = \vec{b}$$

But we want the best solution we can get by projecting ***b*** over the column space, therefore, transforming our system like this:

$$A\hat{x} = \vec{p}$$

Now that we have this, we need to find a way to project vector ***b*** into the column space for us to get vector ***p***. It is to be noted that as vector ***e*** is perpendicular to the plane, then vector ***e*** lives in the orthogonal complement of the column space. The column space of our plane will be denoted by vectors $\boldsymbol{a_1}$ and $\boldsymbol{a_2}$. However, to make an analytical approach we need to expand this to higher dimensional spaces, so we will suppose that we have a vector space with a given basis composed by vectors $\boldsymbol{a_1}$, $\boldsymbol{a_2}$ ... and $\boldsymbol{a_m}$ (each vector with ***n*** components). From this conditions we know that our error vector must fulfill that:

$$\overrightarrow{a_1}\cdot\vec{e} = 0 \rightarrow a_{11}\cdot e_1 + a_{21}\cdot e_2 + \ldots + a_{n1}\cdot e_n = 0$$

$$\overrightarrow{a_2}\cdot\vec{e} = 0 \rightarrow a_{12}\cdot e_1 + a_{22}\cdot e_2 + \ldots + a_{n2}\cdot e_n = 0$$

$$\vdots$$

$$\overrightarrow{a_m}\cdot\vec{e} = 0 \rightarrow a_{1m}\cdot e_1 + a_{2m}\cdot e_2 + \ldots + a_{nm}\cdot e_n = 0$$

Note that this $\boldsymbol{a_i}$ column vectors are the ones that are in the coefficient matrix of the system of equations:

$$A = \begin{pmatrix} a_{11} & a_{12} & \cdots & a_{1m} \\ a_{21} & a_{22} & \cdots & a_{2m} \\ \vdots & \vdots & \cdots & \vdots \\ a_{n1} & a_{n2} & \cdots & a_{nm} \end{pmatrix} = \left(\vec{a_1}, \vec{a_2}, \ldots, \vec{a_n} \right)$$

Therefore, our system can be brought in matrix notation like this:

$$A^T \cdot \vec{e} = \vec{0}$$

Substituting the expression for vector ***e*** and vector ***p***, we get:

$$A^T \cdot \left(\vec{b} - \vec{p} \right) = \vec{0}$$

$$A^T \cdot \left(\vec{b} - A\hat{x} \right) = \vec{0}$$

$$\hat{x} = \left(A^T \cdot A \right)^{-1} \cdot A^T \vec{b} \text{ ,}$$

Which is exactly what we did with the example we had before. This is a procedure that finds the best solution to the system of equations we could get by projecting the independent term into the column space and minimizing the error. This is called the ***Least Squares Method***.

An interesting fact about this is that we found new type of matrices, which are called ***projection matrices***. We found a matrix that projects vector ***b*** into the column space. Let us call the projection matrix ***P***, therefore,:

$$\vec{p} = A\hat{x} = \left(A \cdot \left(A^T \cdot A \right)^{-1} \cdot A^T \right) \vec{b} = P\vec{b}$$

$$P = A \cdot \left(A^T \cdot A \right)^{-1} \cdot A^T$$

But what would happen if we project vector ***p*** again into the column space? We should get the same vector ***p***, since ***p*** is already into the column space. Let us check that:

$$P\vec{p} = P^2\vec{b} = \left(A\cdot\left(A^T\cdot A\right)^{-1}\cdot A^T\right)\left(A\cdot\left(A^T\cdot A\right)^{-1}\cdot A^T\right)\vec{b}$$

$$P^2\vec{b} = A\cdot\left(A^T\cdot A\right)^{-1}\cdot\left(A^T\cdot A\right)\cdot\left(A^T\cdot A\right)^{-1}\cdot A^T\vec{b}$$

$$P^2\vec{b} = A\cdot\left(A^T\cdot A\right)^{-1}\cdot A^T\vec{b} = P\vec{b}$$

$$P\vec{p} = \vec{p}$$

Therefore, we are right -- the projection of vector ***p*** over the column space is vector ***p*** since it was already in the column space. In linear algebra everything has to make sense. Now we can see there are some matrices that share some special properties. For example, projection matrices have the following property:

$$P = P^2 = P^3 = \ldots = P^n$$

Other matrices that share a very important property that will be very useful in other subjects are orthogonal matrices.

ORTHOGONAL MATRICES

Orthogonal matrices are a special type of matrices in which all the columns of the matrix are orthogonal to each other. This is a very important property that is desirable in many cases, but before discussing it, let us check some examples of orthogonal matrices:

$$I = \begin{pmatrix} 1 & 0 & 0 \\ 0 & 1 & 0 \\ 0 & 0 & 1 \end{pmatrix}; A = \begin{pmatrix} 1 & 0 \\ 0 & 1 \\ 1 & 0 \end{pmatrix}; B = \begin{pmatrix} \cos\theta & -\sin\theta & 0 \\ \sin\theta & \cos\theta & 0 \\ 0 & 0 & 1 \end{pmatrix}$$

These are examples of orthogonal matrices since all of their columns are orthogonal to each other. But an important property of them comes better when the columns of this matrices are not only orthogonal but also orthonormal. That is because if the columns of this matrices are orthonormal, then their inverses are their transposes. That is,

$$A^{-1} = A^T$$

In the examples shown above, matrix B is an orthonormal matrix, therefore, its transpose is equal to its inverse:

$$B \cdot B^T = I = \begin{pmatrix} \cos\theta & -\sin\theta & 0 \\ \sin\theta & \cos\theta & 0 \\ 0 & 0 & 1 \end{pmatrix} \cdot \begin{pmatrix} \cos\theta & \sin\theta & 0 \\ -\sin\theta & \cos\theta & 0 \\ 0 & 0 & 1 \end{pmatrix} = \begin{pmatrix} 1 & 0 & 0 \\ 0 & 1 & 0 \\ 0 & 0 & 1 \end{pmatrix}$$

Why this happens is easy to show. Suppose we have a generic matrix ***A*** formed by vectors $\boldsymbol{a}_1$, $\boldsymbol{a}_2$,…, $\boldsymbol{a}_m$, and suppose that these vectors are a set of orthonormal vectors, then:

$$A = \left(\vec{a_1}, \vec{a_2}, \ldots, \vec{a_m}\right); A^T = \begin{pmatrix} \left(\vec{a_1}\right)^T \\ \left(\vec{a_2}\right)^T \\ \vdots \\ \left(\vec{a_m}\right)^T \end{pmatrix}$$

wherein if we assume each vector has ***n*** components then both matrices would look like the generic matrices we have seen. The interesting part is that if we multiply ***A*** by its transpose we get:

$$A^T \cdot A = \begin{pmatrix} \left(\vec{a_1} \cdot \vec{a_1}\right) & \left(\vec{a_1} \cdot \vec{a_2}\right) & \ldots & \left(\vec{a_1} \cdot \vec{a_m}\right) \\ \left(\vec{a_2} \cdot \vec{a_1}\right) & \left(\vec{a_2} \cdot \vec{a_2}\right) & \ldots & \left(\vec{a_2} \cdot \vec{a_m}\right) \\ \vdots & \vdots & \ldots & \vdots \\ \left(\vec{a_m} \cdot \vec{a_1}\right) & \left(\vec{a_m} \cdot \vec{a_2}\right) & \ldots & \left(\vec{a_m} \cdot \vec{a_m}\right) \end{pmatrix}$$

wherein each pair of parentheses represents a component of the new matrix and each component is a dot product between two vectors. From the condition that all of the column vectors in the matrix were orthonormal then all the dot products in the matrix are zero except for those who are at the main diagonal. The dot products that are at the main diagonal will be the dot products of a vector and itself. Therefore, they represent the magnitude of those vectors. But since they are unit vectors, their result is 1, and therefore,

$$A^T \cdot A = I = \begin{pmatrix} 1 & 0 & \dots & 0 \\ 0 & 1 & \dots & 0 \\ \vdots & \vdots & \vdots & \vdots \\ 0 & 0 & \dots & 1 \end{pmatrix}$$

Another fact about this matrices is that orthogonal matrices can only have less or equal columns than rows, but not more columns than rows. The reader should be able to tell why (***hint:*** more columns than rows implies linear dependency on the columns). By the way, if the vectors of matrix $\boldsymbol{A}$ would have not been orthonormal but just orthogonal, the result of multiplying $\boldsymbol{A}$ by its transpose would be a diagonal matrix.

Having special matrices is a blessing when working with them since they present special properties that allow the user to manipulate the system to its benefit. For example, projection matrices are useful because their products will be all the same matrix P, and orthogonal matrices, which are better denoted by the letter Q, have the advantage of an easy inverse. Many more type of matrices exist. Let us check them out.

An important detail before starting this point is that, even when the initial matrix does not have any special property, they are arrangements that are well based and easy to compute to get our matrix into a "special" matrix. These methods are called ***matrix decomposition or matrix factorization.*** Countless matrix factorization exists today, but each of them depends on the need. However, sometimes many of them are the same but have other names. Here we will see the most common ones and the ones that may serve well to solve the vast majority of the problems.

LU Factorization

LU factorization is an arrangement done to square matrices, in which we split them into a multiplication between a ***lower triangular matrix*** and an ***upper triangular matrix***. This arrangement has certain properties and advantages with respect to having a bigger matrix. For example, calculating the determinant of two triangular matrices is a lot easier than calculating the determinant of a full matrix (we will get to this later). Besides, it sometimes represents an advantage when solving system of equations. For example, suppose we have our generic system:

$$A\vec{x} = \vec{b}$$

With ***A*** being a matrix with non zero coefficients. If we are able to split ***A*** into the following form:

$$A = \begin{pmatrix} a_{11} & a_{12} & \cdots & a_{1n} \\ a_{21} & a_{22} & \cdots & a_{2n} \\ \vdots & \vdots & \cdots & \vdots \\ a_{n1} & a_{n2} & \cdots & a_{nn} \end{pmatrix} = L \cdot U = \begin{pmatrix} l_{11} & 0 & \cdots & 0 \\ l_{21} & l_{22} & \cdots & 0 \\ \vdots & \vdots & \cdots & \vdots \\ l_{n1} & l_{n2} & \cdots & l_{nn} \end{pmatrix} \cdot \begin{pmatrix} u_{11} & u_{12} & \cdots & u_{1n} \\ 0 & u_{22} & \cdots & u_{2n} \\ \vdots & \vdots & \cdots & \vdots \\ 0 & 0 & \cdots & u_{nn} \end{pmatrix}$$

Then we could also split our system of equations like this:

$$L \cdot U\vec{x} = \vec{b}$$

$$L\vec{z} = \vec{b}$$

$$U\vec{x} = \vec{z}$$

Which represents an equivalent system of equations. This gives us the advantage that we could use forward and backward substitution to solve for vector ***x***, because we have two triangular matrices. Triangular matrices are very useful in solving systems of equations, since each row of the matrix will give the value of one variable, making them really straightforward to use. However, sometimes finding a factorization for ***A*** can be harder than solving the original system. Therefore, there are some conditions that make this method better depending on the occasion.

In case we want to factor ***A*** into two smaller matrices, we would encounter a problem -- we will have more coefficients than equations, and therefore, the solution for this problem is to give a number to certain coefficients of ***U*** and ***L***. Values are given depending on the method used and author, for example, firstly we have ***Doolittle factorization*** method.

Doolittle factorization method consists of splitting ***A*** into two triangular matrices but ***the lower triangular matrix has all elements in the main diagonal equal to 1***. Therefore, it would look like this:

$$A = \begin{pmatrix} a_{11} & a_{12} & \cdots & a_{1n} \\ a_{21} & a_{22} & \cdots & a_{2n} \\ \vdots & \vdots & \cdots & \vdots \\ a_{n1} & a_{n2} & \cdots & a_{nn} \end{pmatrix} = L \cdot U = \begin{pmatrix} 1 & 0 & \cdots & 0 \\ l_{21} & 1 & \cdots & 0 \\ \vdots & \vdots & \cdots & \vdots \\ l_{n1} & l_{n2} & \cdots & 1 \end{pmatrix} . \begin{pmatrix} u_{11} & u_{12} & \cdots & u_{1n} \\ 0 & u_{22} & \cdots & u_{2n} \\ \vdots & \vdots & \cdots & \vdots \\ 0 & 0 & \cdots & u_{nn} \end{pmatrix}$$

This gives equal number of variables and equations. Therefore, matrix ***A*** has a single Doolittle factorization. An example will be given after explaining the second method which is ***Crout's factorization*** method, the opposite of Doolittle since at this time ***the upper triangular matrix that has all elements in the main diagonal is equal to 1***. Therefore, Crout's transformation would look like the following:

$$A = \begin{pmatrix} a_{11} & a_{12} & \cdots & a_{1n} \\ a_{21} & a_{22} & \cdots & a_{2n} \\ \vdots & \vdots & \cdots & \vdots \\ a_{n1} & a_{n2} & \cdots & a_{nn} \end{pmatrix} = L \cdot U = \begin{pmatrix} l_{11} & 0 & \cdots & 0 \\ l_{21} & l_{22} & \cdots & 0 \\ \vdots & \vdots & \cdots & \vdots \\ l_{n1} & l_{n2} & \cdots & l_{nn} \end{pmatrix} . \begin{pmatrix} 1 & u_{12} & \cdots & u_{1n} \\ 0 & 1 & \cdots & u_{2n} \\ \vdots & \vdots & \cdots & \vdots \\ 0 & 0 & \cdots & 1 \end{pmatrix}$$

Some methods also transform ***A*** into an ***LDU transformation*** that combines both methods like the following way:

$$A = \begin{pmatrix} a_{11} & a_{12} & \cdots & a_{1n} \\ a_{21} & a_{22} & \cdots & a_{2n} \\ \vdots & \vdots & \cdots & \vdots \\ a_{n1} & a_{n2} & \cdots & a_{nn} \end{pmatrix} = L \cdot D \cdot U = \begin{pmatrix} 1 & 0 & \cdots & 0 \\ l_{21} & 1 & \cdots & 0 \\ \vdots & \vdots & \cdots & \vdots \\ l_{n1} & l_{n2} & \cdots & 1 \end{pmatrix} . \begin{pmatrix} d_1 & 0 & \cdots & 0 \\ 0 & d_2 & \cdots & 0 \\ \vdots & \vdots & \cdots & \vdots \\ 0 & 0 & \cdots & d_n \end{pmatrix} . \begin{pmatrix} 1 & u_{12} & \cdots & u_{1n} \\ 0 & 1 & \cdots & u_{2n} \\ \vdots & \vdots & \cdots & \vdots \\ 0 & 0 & \cdots & 1 \end{pmatrix}$$

However, all these methods are similar and equivalent since they should all be equal to matrix ***A***. An important detail is that these methods require that the matrix we are factoring is full rank, because if it is not full rank then it would not have a single factorization and we would have more variables than equations. That is, ***A must be full rank***, and a consequence of this is that ***A must also be square.*** Having said this, let us do some example and see the differences among the three methods.

Suppose we have a matrix ***A*** like the one below, and we want to factor it:

$$A = \begin{pmatrix} 2 & 4 & 6 \\ 4 & 5 & 6 \\ 3 & 1 & -2 \end{pmatrix}$$

Matrix A is full rank since it has independent columns and its determinant is non zero (its determinant is 6) and is also a square, and therefore, it admits ***LU*** decomposition. Its ***Doolittle factorization*** would look like this:

$$A = \begin{pmatrix} 2 & 4 & 6 \\ 4 & 5 & 6 \\ 3 & 1 & -2 \end{pmatrix} = \begin{pmatrix} 1 & 0 & 0 \\ 2 & 1 & 0 \\ 3/2 & 5/3 & 1 \end{pmatrix} \cdot \begin{pmatrix} 2 & 4 & 6 \\ 0 & -3 & -6 \\ 0 & 0 & -1 \end{pmatrix} = L \cdot U$$

Its ***Crout's factorization*** would be:

$$A = \begin{pmatrix} 2 & 4 & 6 \\ 4 & 5 & 6 \\ 3 & 1 & -2 \end{pmatrix} = \begin{pmatrix} 2 & 0 & 0 \\ 4 & -3 & 0 \\ 3 & -5 & -1 \end{pmatrix} \cdot \begin{pmatrix} 1 & 2 & 3 \\ 0 & 1 & 2 \\ 0 & 0 & 1 \end{pmatrix} = L \cdot U$$

And its ***LDU factorization*** is:

$$A = \begin{pmatrix} 2 & 4 & 6 \\ 4 & 5 & 6 \\ 3 & 1 & -2 \end{pmatrix} = \begin{pmatrix} 1 & 0 & 0 \\ 2 & 1 & 0 \\ 3/2 & 5/3 & 1 \end{pmatrix} \cdot \begin{pmatrix} 2 & 0 & 0 \\ 0 & -3 & 0 \\ 0 & 0 & -1 \end{pmatrix} \cdot \begin{pmatrix} 1 & 2 & 3 \\ 0 & 1 & 2 \\ 0 & 0 & 1 \end{pmatrix} = L \cdot D \cdot U$$

As you can see, an ***LDU factorization*** is more general and it combines both methods. Doing ***LU*** factorizations is quite simple since each coefficient or variable is calculated on each equation. To see how these methods can be convenient, let us suppose we have the following system of equations:

$$\begin{pmatrix} 2 & 4 & 6 \\ 4 & 5 & 6 \\ 3 & 1 & -2 \end{pmatrix} \begin{pmatrix} x_1 \\ x_2 \\ x_3 \end{pmatrix} = \begin{pmatrix} 18 \\ 24 \\ 4 \end{pmatrix}$$

If we factor the coefficient matrix into an ***LU*** form, suppose Doolittle form, we get:

$$\begin{pmatrix} 1 & 0 & 0 \\ 2 & 1 & 0 \\ 3/2 & 5/3 & 1 \end{pmatrix} \cdot \begin{pmatrix} 2 & 4 & 6 \\ 0 & -3 & -6 \\ 0 & 0 & -1 \end{pmatrix} \begin{pmatrix} x_1 \\ x_2 \\ x_3 \end{pmatrix} = \begin{pmatrix} 18 \\ 24 \\ 4 \end{pmatrix}$$

And we could transform the above system into two systems:

$$\begin{pmatrix} 1 & 0 & 0 \\ 2 & 1 & 0 \\ 3/2 & 5/3 & 1 \end{pmatrix} \cdot \begin{pmatrix} z_1 \\ z_2 \\ z_3 \end{pmatrix} = \begin{pmatrix} 18 \\ 24 \\ 4 \end{pmatrix}$$

$$\begin{pmatrix} 2 & 4 & 6 \\ 0 & -3 & -6 \\ 0 & 0 & -1 \end{pmatrix} \begin{pmatrix} x_1 \\ x_2 \\ x_3 \end{pmatrix} = \begin{pmatrix} z_1 \\ z_2 \\ z_3 \end{pmatrix}$$

From here we could first find the z vector, and then find the x vector, with front and back substitution. From the first system we get that vector z is:

$$\begin{pmatrix} z_1 \\ z_2 \\ z_3 \end{pmatrix} = \begin{pmatrix} 18 \\ -12 \\ -3 \end{pmatrix}$$

Therefore, from the second system of equations we get that vector $\boldsymbol{x}$ is:

$$\begin{pmatrix} x_1 \\ x_2 \\ x_3 \end{pmatrix} = \begin{pmatrix} 4 \\ -2 \\ 3 \end{pmatrix}$$

Now that we have seen this, let us check one special factorization that is done on symmetrical matrices, and that is ***Cholesky's factorization***. Cholesky's factorization is a special ***LU*** factorization in which the lower triangular matrix is the transpose of the upper triangular matrix. Therefore,

$$A = L \cdot U = U^T \cdot U = L \cdot L^T$$

This comes from the fact that a matrix times its transpose is assured to be a symmetrical matrix. However, this factorization is not always possible. There are some restrictions aside from the fact that $\boldsymbol{A}$ is a square matrix and full rank. *For a symmetric matrix to have a Cholesky's factorization, the matrix needs to be* ***positive definite****.*

SHORT PARENTHESIS: POSITIVE DEFINITE MATRICES

The usual definition of positive definite matrices is the following: Given a square matrix $\boldsymbol{A}$ *and an arbitrary vector* $\boldsymbol{x}$*, then* $\boldsymbol{A}$ *is positive definite if:*

$$\overrightarrow{x^T} \cdot A \cdot \vec{x} > 0$$

*For any vector **x** different than zero.*

*Let us compute a form for the above statement, suppose matrix **A** and vector **x** have the following form:*

$$A = \begin{pmatrix} a_{11} & a_{12} & \cdots & a_{1n} \\ a_{21} & a_{22} & \cdots & a_{2n} \\ \vdots & \vdots & \cdots & \vdots \\ a_{n1} & a_{n2} & \cdots & a_{nn} \end{pmatrix}; \vec{x} = \begin{pmatrix} x_1 \\ x_2 \\ \vdots \\ x_n \end{pmatrix}$$

Then:

$$A \cdot \vec{x} = \begin{pmatrix} a_{11} & a_{12} & \cdots & a_{1n} \\ a_{21} & a_{22} & \cdots & a_{2n} \\ \vdots & \vdots & \cdots & \vdots \\ a_{n1} & a_{n2} & \cdots & a_{nn} \end{pmatrix} \cdot \begin{pmatrix} x_1 \\ x_2 \\ \vdots \\ x_n \end{pmatrix} = \begin{pmatrix} a_{11} \cdot x_1 + a_{12} \cdot x_2 + \cdots + a_{1n} \cdot x_n \\ a_{21} \cdot x_1 + a_{22} \cdot x_2 + \cdots + a_{2n} \cdot x_n \\ \vdots \\ a_{n1} \cdot x_1 + a_{n2} \cdot x_2 + \cdots + a_{nn} \cdot x_n \end{pmatrix}$$

And:

$$x^T \cdot A \cdot \vec{x} = (x_1 \quad x_2 \quad \cdots \quad x_n) \cdot \begin{pmatrix} a_{11} \cdot x_1 + a_{12} \cdot x_2 + \cdots + a_{1n} \cdot x_n \\ a_{21} \cdot x_1 + a_{22} \cdot x_2 + \cdots + a_{2n} \cdot x_n \\ \vdots \\ a_{n1} \cdot x_1 + a_{n2} \cdot x_2 + \cdots + a_{nn} \cdot x_n \end{pmatrix}$$

$$= \begin{matrix} (a_{11}x_1^2 + a_{12}x_1x_2 + \ldots + a_{1n}x_1x_n) \\ +(a_{21}x_2x_1 + a_{22}x_2^2 + \ldots + a_{2n}x_2x_n) \\ + \cdots \\ +(a_{n1}x_nx_1 + a_{n2}x_nx_2 + \ldots + a_{nn}x_n^2) \end{matrix}$$

*Therefore, the result is a scalar, or a number that must be greater than zero for matrix **A** to be positive definite. The above can be expressed like*

$$x^T \cdot A \cdot x = \sum_{i=1}^{n} \sum_{j=1}^{n} a_{ij} \cdot x_i \cdot x_j$$

This expression may be difficult to work with. However, many other methods can be used to check if a matrix is positive definite. These methods are equivalent and if you think about it, they all make sense. These methods are:

- Matrix A must be full rank. If not, then there can be a way such that the above expression may equal 0.
- Matrix A must have the elements in the diagonal positive when applying row reduction.
- The above expression is also satisfied when the upper left determinants are positive (this will be explained soon; this condition comes from Sylvester's criterion).
- Also, matrix A must have all positive eigenvalues (it will be discussed later in another chapter).

However, there is one condition that is easier to check and fulfills all of the other conditions. This condition is that "matrix A must be diagonally dominant." This means that all coefficients in the main diagonal of A are equal or greater than the sum of the absolute values of the elements in their rows. Therefore, if A is a diagonally dominant,

$$a_{ii} \geq \sum_{\substack{i=1 \\ i \neq j}}^{n} \left| a_{ij} \right|$$

*However, the opposite is not true because if we know that **A** is definite positive, it does not assure that **A** is diagonally dominant.*

Before getting into Cholesky's factorization, let us explain these properties with some examples. Let us use one of the matrices we have seen and check if it is positive definite:

$$A = \begin{pmatrix} 2 & 4 & 6 \\ 4 & 5 & 6 \\ 3 & 1 & -2 \end{pmatrix}$$

Therefore,

$$x^T \cdot A \cdot x = 2x_1^2 + 8x_1x_2 + 9x_1x_3 + 5x_2^2 + 7x_2x_3 - 2x_3^2$$

which is not positive for any vector x. For example, if vector x is (-1, 1, 1) then the product is negative. Therefore, the above matrix did not accomplish some of the properties we saw before:

- The above matrix is not diagonally dominant, therefore, we could suspect that it may not be positive definite.

- If we do row reduction on A (let us call the reduced matrix A') we notice that the elements in the diagonal are not positive:

$$A' = \begin{pmatrix} 2 & 4 & 6 \\ 0 & -3 & -6 \\ 0 & 0 & -1 \end{pmatrix}$$

- In this case, however, matrix A is full rank.
- And the upper left determinants are not positive. Let us see what this means. The upper left determinants are all the determinants that start in the upper left side of the matrix and grow one row and column at a time. Using matrix A as example we get:

$$det(A_1) = |2| = 2; det(A_2) = \begin{vmatrix} 2 & 4 \\ 4 & 5 \end{vmatrix} = -6; det(A_3) = \begin{vmatrix} 2 & 4 & 6 \\ 4 & 5 & 6 \\ 3 & 1 & -2 \end{vmatrix} = 6$$

We see that not all of the determinants are positive, therefore, not positive definite. Now let us look at ***an example that does fulfill*** *all the properties:*

$$B = \begin{pmatrix} 2 & -1 & 0 \\ -1 & 2 & -1 \\ 0 & -1 & 2 \end{pmatrix}$$

Therefore,

$$x^T \cdot B \cdot x = 2x_1^2 - 2x_1x_2 + 2x_2^2 - 2x_2x_3 + 2x_3^2$$

$$x^T \cdot B \cdot x = x_1^2 + (x_1 - x_2)^2 + (x_2 - x_3)^2 + x_3^2$$

which is always positive for every vector $\mathbf{x}$*. However, we will prove the other properties:*

- The above matrix is diagonally dominant.
- Doing row reduction we get:

$$B' = \begin{pmatrix} 2 & -1 & 0 \\ 0 & 3/2 & -1 \\ 0 & 0 & 4/3 \end{pmatrix}$$

We can see that all the elements in the main diagonal are positive.

- The matrix is full rank.
- The upper left determinants are all positive:

$$det(A_1) = |2| = 2; det(A_2) = \begin{vmatrix} 2 & -1 \\ -1 & 2 \end{vmatrix} = 3; det(A_3) = \begin{vmatrix} 2 & -1 & 0 \\ -1 & 2 & -1 \\ 0 & -1 & 2 \end{vmatrix} = 4$$

Now that we saw this, we can return to our case of study which is ***Cholesky's factorization***. The only condition for a symmetric matrix to have a Cholesky's decomposition is that the matrix is positive definite. However, if the matrix is diagonally dominant then that condition is already accomplished.

We have already seen how we can do an ***LU*** decomposition, however, Cholesky's decomposition is more desirable because it reduces the number of variables and is more stable while solving systems of equations. Let us look at the following matrix for example:

$$A = \begin{pmatrix} 1 & -1 & 1 \\ -1 & 5 & 1 \\ 1 & 1 & 3 \end{pmatrix}$$

which is not diagonally dominant, but it is symmetric. Let us look at the properties we saw before. First, let us reduce the matrix to its echelon form. In that way we can check all of the properties because we can check the elements in the diagonal, the upper left determinants and the rank. Reducing the matrix we get

$$A' = \begin{pmatrix} 1 & -1 & 1 \\ 0 & 4 & 2 \\ 0 & 0 & 1 \end{pmatrix}$$

From this reduction we can see that the matrix is full rank, all of the elements in the reduced matrix are positive, and we can check the upper left determinants (doing row reduction does not affect the determinant. We will show why later on):

$$det(A_1) = |1| = 1; det(A_2) = \begin{vmatrix} 1 & -1 \\ 0 & 4 \end{vmatrix} = 4; det(A_3) = \begin{vmatrix} 1 & -1 & 1 \\ 0 & 4 & 2 \\ 0 & 0 & 1 \end{vmatrix} = 4$$

Therefore, matrix A is definite positive, and so it has a Cholesky's decomposition. Let us look for that:

$$A = \begin{pmatrix} 1 & -1 & 1 \\ -1 & 5 & 1 \\ 1 & 1 & 3 \end{pmatrix} = \begin{pmatrix} l_{11} & 0 & 0 \\ l_{21} & l_{22} & 0 \\ l_{31} & l_{32} & l_{33} \end{pmatrix} \cdot \begin{pmatrix} l_{11} & l_{21} & l_{31} \\ 0 & l_{22} & l_{32} \\ 0 & 0 & l_{33} \end{pmatrix} = L \cdot L^T$$

Therefore, solving the system we get:

$$A = \begin{pmatrix} 1 & -1 & 1 \\ -1 & 5 & 1 \\ 1 & 1 & 3 \end{pmatrix} = \begin{pmatrix} 1 & 0 & 0 \\ -1 & 2 & 0 \\ 1 & 1 & -1 \end{pmatrix} \cdot \begin{pmatrix} 1 & -1 & 1 \\ 0 & 2 & 1 \\ 0 & 0 & -1 \end{pmatrix}$$

That is Cholesky's factorization for the original matrix. However, it's important to mention that there exists another factorization:

$$A = \begin{pmatrix} 1 & -1 & 1 \\ -1 & 5 & 1 \\ 1 & 1 & 3 \end{pmatrix} = \begin{pmatrix} -1 & 0 & 0 \\ 1 & 2 & 0 \\ -1 & 1 & 1 \end{pmatrix} \cdot \begin{pmatrix} -1 & 1 & -1 \\ 0 & 2 & 1 \\ 0 & 0 & 1 \end{pmatrix}$$

which is not very different from the one we had before. However, this can happen in some cases. This covers the LU decomposition, but many other decompositions are possible.

QR DECOMPOSITION

QR decomposition is a decomposition possible on any square matrix, which is done by decomposing a matrix into an orthonormal matrix $\boldsymbol{Q}$ and an upper triangular matrix, usually denoted by the letter $\boldsymbol{R}$. This method represents the advantage in which the inverse of matrix ***Q*** is its transpose. This method is used in many applications. You will see some of them in other chapters but one of them is the following. Suppose we have a matrix ***A*** that has a ***QR*** decomposition:

$$A = Q \cdot R$$

The inverse of ***A*** is easily computed:

$$A^{-1} = R^{-1} Q^{T}$$

Since the inverse of ***Q*** is its transpose and inverting an upper triangular matrix is also easy. This process of ***QR*** decomposition is possible in any square matrix because of the following. If we have a matrix ***A*** of the form:

$$A = \begin{pmatrix} a_{11} & a_{12} & \cdots & a_{1n} \\ a_{21} & a_{22} & \cdots & a_{2n} \\ \vdots & \vdots & \cdots & \vdots \\ a_{n1} & a_{n2} & \cdots & a_{nn} \end{pmatrix}$$

We know that each column vector in ***A*** is a vector that is part of a basis of the column space, therefore, we could orthonormalize the vectors in matrix ***A***, and then we could find another matrix that combines the orthonormal vectors to get ***A*** again. This can be better seen using Gram Schmidt like the following. Suppose each column vector in the matrix is denoted like this:

$$A = \left(\vec{a_1}, \vec{a_2}, \ldots, \vec{a_n}\right) \rightarrow \vec{a_1} = \begin{pmatrix} a_{11} \\ a_{21} \\ \vdots \\ a_{n1} \end{pmatrix}; \vec{a_2} = \begin{pmatrix} a_{12} \\ a_{22} \\ \vdots \\ a_{n2} \end{pmatrix}; \ldots; \vec{a_n} = \begin{pmatrix} a_{1n} \\ a_{2n} \\ \vdots \\ a_{nn} \end{pmatrix};$$

If we apply Gram Schmidt starting with vector and finalizing with vector , then we would obtain the following equations:

$$\vec{u_1}' = \vec{a_1} \rightarrow \vec{u_1} = \frac{\vec{u_1}'}{\|\vec{u_1}'\|} \qquad \vec{a_1} = \|\vec{u_1}'\| \cdot \vec{u_1}$$

$$\vec{u_2}' = \vec{a_2} - (\vec{a_2} \cdot \vec{u_1})\vec{u_1} \rightarrow \vec{u_2} = \frac{\vec{u_2}'}{\|\vec{u_2}'\|} \qquad \vec{a_2} = \|\vec{u_2}'\| \cdot \vec{u_2} + (\vec{a_2} \cdot \vec{u_1})\vec{u_1}$$

$$\vec{u_3}' = \vec{a_3} - (\vec{a_3} \cdot \vec{u_1})\vec{u_1} - (\vec{a_3} \cdot \vec{u_2})\vec{u_2} \rightarrow \vec{u_3} = \frac{\vec{u_3}'}{\|\vec{u_3}'\|} \qquad \vec{a_3} = \|\vec{u_3}'\| \cdot \vec{u_3} + (\vec{a_3} \cdot \vec{u_1})\vec{u_1} + (\vec{a_3} \cdot \vec{u_2})\vec{u_2}$$

$$\vdots \qquad \vdots$$

$$\vec{u_n}' = \vec{a_n} - \sum_{i=1}^{n-1} (\vec{a_n} \cdot \vec{u_i})\,\vec{u_i} \rightarrow \vec{u_n} = \frac{\vec{u_n}'}{\|\vec{u_n}'\|} \qquad \vec{a_n} = \|\vec{u_n}'\| \cdot \vec{u_n} + \sum_{i=1}^{n-1} (\vec{a_n} \cdot \vec{u_i})\,\vec{u_i}$$

This can be expressed under matrix notation like:

$$(\vec{a_1}, \vec{a_2}, \vec{a_3}, \ldots, \vec{a_n}) = (\vec{u_1}, \vec{u_2}, \vec{u_3}, \ldots, \vec{u_n}) \cdot \begin{pmatrix} \|\vec{u_1}'\| & (\vec{a_2} \cdot \vec{u_1}) & (\vec{a_3} \cdot \vec{u_1}) & \cdots & (\vec{a_n} \cdot \vec{u_1}) \\ 0 & \|\vec{u_2}'\| & (\vec{a_3} \cdot \vec{u_2}) & \cdots & (\vec{a_n} \cdot \vec{u_2}) \\ 0 & 0 & \|\vec{u_3}'\| & \cdots & (\vec{a_n} \cdot \vec{u_3}) \\ \vdots & \vdots & \vdots & \cdots & \vdots \\ 0 & 0 & 0 & \cdots & \|\vec{u_n}'\| \end{pmatrix}$$

Where we can see that matrix A can be rewritten as a multiplication of an orthonormal matrix and an upper triangular matrix, which proves the algorithm. However, to calculate this, we usually orthonormalize the columns of A and calculate matrix R like this:

$$R = Q^T \cdot A$$

Now that we have seen that this is possible for every matrix, we ask ourselves how to compute these matrices. Actually, we can use all three methods. The first one is to use ***Gram Schmidt orthonormalization*** process, which we already know, the other is to use ***Householder reflections*** and the last one is to use ***Givens rotations***.

Let us make an example from this process to show how it is done. Suppose we have the following matrix:

$$A = \begin{pmatrix} 2 & 4 & 6 \\ 4 & 5 & 6 \\ 3 & 1 & -2 \end{pmatrix}$$

The column vectors in our matrix are:

$$\overrightarrow{a_1} = \begin{pmatrix} 2 \\ 4 \\ 3 \end{pmatrix}; \overrightarrow{a_2} = \begin{pmatrix} 4 \\ 5 \\ 1 \end{pmatrix}; \overrightarrow{a_3} = \begin{pmatrix} 6 \\ 6 \\ -2 \end{pmatrix}$$

For our ***QR*** decomposition to work, it is important to start with the column vectors from left to right, that is, we need to start with vector and finish with. If we apply Gram Schmidt in this example, then we can start orthonormalizing the set of vectors. In this case, since matrix ***Q*** is denoted by the letter ***Q***, we will denote the orthonormal vectors liketo have a better notation, therefore,

$$\overrightarrow{a_1} = \begin{pmatrix} 2 \\ 4 \\ 3 \end{pmatrix}; \overrightarrow{q_1} = \frac{1}{\sqrt{29}} \begin{pmatrix} 2 \\ 4 \\ 3 \end{pmatrix}$$

$$\overrightarrow{a_2} = \begin{pmatrix} 4 \\ 5 \\ 1 \end{pmatrix}; \vec{q}\,'_2 = \overrightarrow{a_2} - \left(\overrightarrow{q_1} \cdot \overrightarrow{a_2}\right)\overrightarrow{q_1}$$

$$\vec{q}\,'_2 = \begin{pmatrix} 4 \\ 5 \\ 1 \end{pmatrix} - \frac{(2 \cdot 4 + 4 \cdot 5 + 3 \cdot 1)}{29} \begin{pmatrix} 2 \\ 4 \\ 3 \end{pmatrix} = \begin{pmatrix} 4 \\ 5 \\ 1 \end{pmatrix} - \frac{31}{29} \begin{pmatrix} 2 \\ 4 \\ 3 \end{pmatrix} = \frac{1}{29} \begin{pmatrix} 54 \\ 21 \\ -64 \end{pmatrix}$$

$$\overrightarrow{q_2} = \frac{1}{\sqrt{7453}} \begin{pmatrix} 54 \\ 21 \\ -64 \end{pmatrix}$$

$$\overrightarrow{a_3} = \begin{pmatrix} 6 \\ 6 \\ -2 \end{pmatrix}; \vec{q}\,'_3 = \overrightarrow{a_3} - \left(\overrightarrow{q_1} \cdot \overrightarrow{a_3}\right)\overrightarrow{q_1} - \left(\overrightarrow{q_2} \cdot \overrightarrow{a_3}\right)\overrightarrow{q_2}$$

$$\vec{q}\,'_3 = \begin{pmatrix} 6 \\ 6 \\ -2 \end{pmatrix} - \frac{30}{29} \begin{pmatrix} 2 \\ 4 \\ 3 \end{pmatrix} - \frac{578}{7453} \begin{pmatrix} 54 \\ 21 \\ -64 \end{pmatrix} = \frac{1}{7453} \begin{pmatrix} -1914 \\ 1740 \\ -1044 \end{pmatrix}$$

$$\vec{q_3} = \frac{1}{\sqrt{257}}\begin{pmatrix} -11 \\ 10 \\ -6 \end{pmatrix}$$

Therefore, our matrix ***Q*** would be:

$$Q = \frac{1}{\sqrt{7453}}\begin{pmatrix} 2\sqrt{257} & 54 & -11\sqrt{29} \\ 4\sqrt{257} & 21 & 10\sqrt{29} \\ 3\sqrt{257} & -64 & -6\sqrt{29} \end{pmatrix}$$

Computing matrix ***R***,

$$R = Q^T \cdot A = Q = \frac{1}{\sqrt{7453}}\begin{pmatrix} 2\sqrt{257} & 4\sqrt{257} & 3\sqrt{257} \\ 54 & 21 & -64 \\ -11\sqrt{29} & 10\sqrt{29} & -6\sqrt{29} \end{pmatrix} \cdot \begin{pmatrix} 2 & 4 & 6 \\ 4 & 5 & 6 \\ 3 & 1 & -2 \end{pmatrix}$$

$$R = \frac{1}{\sqrt{7453}}\begin{pmatrix} 29\sqrt{257} & 31\sqrt{257} & 30\sqrt{257} \\ 0 & 257 & 578 \\ 0 & 0 & 6\sqrt{29} \end{pmatrix}$$

It is true that Gram Schmidt is an amazing algorithm and is easy to implement However, it is not numerically stable since it involves taking square roots that usually cuts decimal places. Also, Gram Schmidt does not behave when two columns of matrix ***A*** are almost parallel, because, if their difference is small, then computers could confuse that difference with a zero. We will see this phenomenon later on.

A much more numerical stable process is Householder reflections, which we will see now with an example of higher size.

HOUSEHOLDER REFLECTION

The householder reflection method is a method in which through successive reflections of the vectors of a matrix it manages to orthogonalize its columns. This method is much more numerically stable than Gram Schmidt and is very intuitive. Let us look at the principle behind how it works.

Suppose we have a vector ***x***, shown in the image below (colored in red) and another vector ***u*** (colored in blue). Suppose we want to reflect vector ***x*** with respect to vector ***v*** (colored in green) in order to get vector ***x'*** (colored in marine blue), it is important to notice that for this to be true, vector ***x, u, v and x'*** must be in the same plane. To do this reflections we do the following approach.

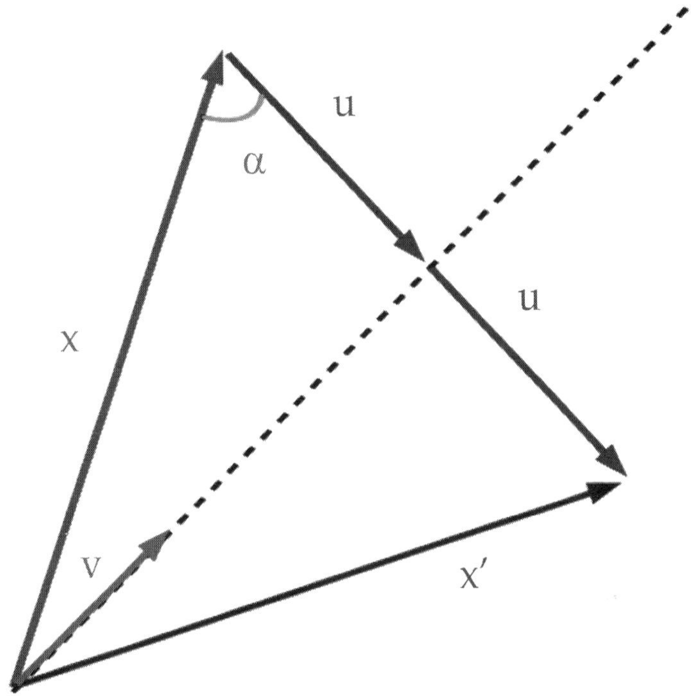

Figure 14: Householder transformation.

First, we need to find a way to correlate vector ***x*** and ***u***. From the image below we may notice that ***u*** is the projection of ***-x*** over ***u***, therefore,:

$$\vec{u} = \frac{-(\vec{x}\cdot\vec{u})}{\| u \|^2}\vec{u}$$

If we want to reflect vector ***x*** with respect to ***v***, then we need to add vector ***x*** and ***2*** times vector ***u***, this will get us to vector ***x'***:

$$\vec{x}' = \vec{x} + 2 \cdot \vec{u}$$

$$\vec{x}' = \vec{x} - 2 \cdot \frac{(\vec{x} \cdot \vec{u})}{\| u \|^2} \vec{u}$$

After some arrangements, it can be rewritten as:

$$\vec{x}' = \left[I - \frac{2 \cdot \vec{u} \overrightarrow{u^T}}{\| u \|^2} \right] \vec{x}$$

You can see that we have a matrix in the above equation. This matrix is called the ***Householder transformation***, which is usually denoted by the letter ***H***, and it reflects a vector ***x*** with respect to a vector ***v*** under the assumption that they are co-planar vectors, however, vector ***v*** does not appear anywhere in the equations. This is because vector ***v*** and vector ***u*** are orthogonal to each other (this is a condition for the reflection). This reflections have special properties, and one of them is that if we reflect vector ***x'*** with respect to vector ***v***, then we would get vector ***x*** again, therefore,:

$$\vec{x} = \left[I - \frac{2 \cdot \vec{u} \overrightarrow{u^T}}{\| u \|^2} \right] \vec{x}' \quad \left[I - \frac{2 \cdot \vec{u} \overrightarrow{u^T}}{\| u \|^2} \right]^2 = I$$

Let us try to prove this.

$$\left[I - \frac{2 \cdot \vec{u} \overrightarrow{u^T}}{\| u \|^2} \right]^2 = I^2 - I \cdot \frac{2 \cdot \vec{u} \overrightarrow{u^T}}{\| u \|^2} - \frac{2 \cdot \vec{u} \overrightarrow{u^T}}{\| u \|^2} \cdot I + \left(\frac{2 \cdot \vec{u} \overrightarrow{u^T}}{\| u \|^2} \right)^2$$

$$\left[I - \frac{2 \cdot \vec{u} \overrightarrow{u^T}}{\| u \|^2} \right]^2 = I - \frac{4 \cdot \vec{u} \overrightarrow{u^T}}{\| u \|^2} + \frac{4 \cdot \left(\vec{u} \overrightarrow{u^T} \right) \cdot \left(\vec{u} \overrightarrow{u^T} \right)}{\| u \|^4}$$

$$\rightarrow \frac{4 \cdot \left(\vec{u} \overrightarrow{u^T} \right) \cdot \left(\vec{u} \overrightarrow{u^T} \right)}{\| u \|^4} = \frac{4 \cdot \vec{u} \left(\overrightarrow{u^T} \cdot \vec{u} \right) \overrightarrow{u^T}}{\| u \|^4} = \frac{4 \cdot \left(\vec{u} \overrightarrow{u^T} \right) \cdot \| u \|^2}{\| u \|^4} = \frac{4 \cdot \vec{u} \overrightarrow{u^T}}{\| u \|^2}$$

$$\left[I - \frac{2\cdot\overrightarrow{\vec{u}u^T}}{\|u\|^2}\right]^2 = I - \frac{4\cdot\overrightarrow{\vec{u}u^T}}{\|u\|^2} + \frac{4\cdot\overrightarrow{\vec{u}u^T}}{\|u\|^2} = I$$

- In this case, like $H^2 = I$, the Householder transformation is called to be involutory. However, it has properties and details that go beyond what we have just seen:
    - ***The first one*** is that it is most useful when we trick this reflections so that vector ***x'*** has the form we want. This will be explained later with an example.
    - ***The second one*** is that, as we are reflecting vector ***x*** into ***x'***, then vector ***x*** and vector ***x'*** have the same length.
    - ***The third one*** is that we could change a little bit of our Householder transformation by making vectors ***u*** and $\boldsymbol{u}^T$ unit vectors. Our Householder transformation would look like this then:

      $$\vec{x}' = \left[I - 2\cdot\widehat{\hat{u}u^T}\right]\vec{x}$$

    - ***The fourth one*** is that as we proved before, $H^2 = I$, therefore, the Householder inverse matrix is itself.

$$H^{-1} = H$$

- ***And the last one*** we are going to mention is that the Householder transformation is also a symmetric matrix:

$$H^T = \left(I - 2\cdot\widehat{\hat{u}u^T}\right)^T = I^T - 2\cdot\left(\widehat{u^T}\right)^T \widehat{u^T} = I - 2\cdot\widehat{\hat{u}u^T} = H$$

As a consequence, the Householder transformation shares all of these properties at the same time and is, therefore, a symmetric, involutory and orthonormal matrix, because:

$$H = H^T = H^{-1}$$

Having this said, let us make an example of high order to see how the Householder transformation can be useful to the ***QR*** decomposition. In this case, we will show all elements of matrices as decimal numbers for two reasons: easement and that in real matrices coefficients are written that way. Let us start with the example. Suppose we have the following matrix ***A*** and we want to find its ***QR*** form:

$$A = \begin{pmatrix} 1 & 2 & 3 & 4 \\ 1 & 3 & 4 & 6 \\ 7 & 8 & 4 & 6 \\ 9 & 7 & 10 & 0 \end{pmatrix}$$

What the Householder algorithm does after several reflections is transform our matrix into an upper diagonal matrix. For this, we take our matrix ***A*** and take its first vector:

$$A = \begin{pmatrix} 1 & 2 & 3 & 4 \\ 1 & 3 & 4 & 6 \\ 7 & 8 & 4 & 6 \\ 9 & 7 & 10 & 0 \end{pmatrix} \to \vec{a_1} = \begin{pmatrix} 1 \\ 1 \\ 7 \\ 9 \end{pmatrix} \to \vec{a_1}' = \begin{pmatrix} \mu \\ 0 \\ 0 \\ 0 \end{pmatrix}$$

If we want to transform A into an upper triangular matrix, then vectormust be transformed into vector, wherein the condition is that both vectors have the same length. In this case, the term will be the length of vector . This will be done using a reflection, that is, we have vectorand we want to reflect it somehow such that we get vector. For this we do the following:

$$\vec{a_1} = \begin{pmatrix} 1 \\ 1 \\ 7 \\ 9 \end{pmatrix}; \| \vec{a_1} \| = \sqrt{132} \to \vec{a_1}' = \begin{pmatrix} 11.489 \\ 0 \\ 0 \\ 0 \end{pmatrix}$$

From here we calculate vector ***u*** which will be the difference between the vector we want to get, that is, $\vec{a_1}'$, and our original vector which is $\vec{a_1}$, and then normalize it, like this:

$$\vec{u} = \overrightarrow{a_1}' - \overrightarrow{a_1} = \begin{pmatrix} 10.489 \\ -1 \\ -7 \\ -9 \end{pmatrix} \rightarrow \hat{u} = \begin{pmatrix} -0.6756 \\ 0.0644 \\ 0.4508 \\ 0.5797 \end{pmatrix}$$

From here, we proceed to calculate the Householder transformation matrix:

$$\hat{u} \cdot \widehat{u^T} = \begin{pmatrix} 0.4564 & -0.0435 & -0.3046 & -0.3916 \\ -0.0435 & 0.0041 & 0.0290 & 0.0373 \\ -0.3046 & 0.029 & 0.2033 & 0.2613 \\ -0.3916 & 0.0373 & 0.2613 & 0.3360 \end{pmatrix}$$

$$I - 2 \cdot \hat{u} \cdot \widehat{u^T} = \begin{pmatrix} 0.0870 & 0.0870 & 0.6092 & 0.7833 \\ 0.0870 & 0.9917 & -0.0580 & -0.0746 \\ 0.6092 & -0.058 & 0.5933 & -0.5227 \\ 0.7833 & -0.0746 & -0.5227 & 0.3278 \end{pmatrix}$$

And that is our Householder transformation which we will name $\boldsymbol{Q_1}$. If we multiply this matrix by our original matrix $\boldsymbol{A}$, we get:

$$Q_1 \cdot A = \begin{pmatrix} 0.0870 & 0.0870 & 0.6092 & 0.7833 \\ 0.0870 & 0.9917 & -0.0580 & -0.0746 \\ 0.6092 & -0.058 & 0.5933 & -0.5227 \\ 0.7833 & -0.0746 & -0.5227 & 0.3278 \end{pmatrix} \cdot \begin{pmatrix} 1 & 2 & 3 & 4 \\ 1 & 3 & 4 & 6 \\ 7 & 8 & 4 & 6 \\ 9 & 7 & 10 & 0 \end{pmatrix}$$

$$Q_1 \cdot A = \begin{pmatrix} 11.489 & 10.792 & 10.879 & 4.526 \\ 0 & 2.1617 & 3.2487 & 5.9498 \\ 0 & 2.1319 & -1.2587 & 5.6489 \\ 0 & -0.5446 & 3.2387 & -0.4513 \end{pmatrix}$$

which is 1 step to make matrix A into an upper diagonal matrix. But to continue making matrix A into a triangular form, we need to do the same process with the other columns. For this we will take the second column, which we call vector. We will transform it through a householder reflection into a vector with the form shown below. Remember that the condition for a Householder transformation is that both vectors and have the same length. Therefore,

$$\vec{a_2} = \begin{pmatrix} 10.792 \\ 2.1617 \\ 2.1319 \\ -0.5446 \end{pmatrix} \rightarrow \vec{a_2}' = \begin{pmatrix} 10.792 \\ \beta \\ 0 \\ 0 \end{pmatrix} \rightarrow \beta = \sqrt{(2.1617)^2 + (2.1319)^2 + (-0.5446)^2} = 3.0845$$

Doing the same procedure as before:

$$\vec{u} = \vec{a_2}' - \vec{a_2} = \begin{pmatrix} 0 \\ -0.9229 \\ 2.1319 \\ -0.5446 \end{pmatrix} \rightarrow \hat{u} = \begin{pmatrix} 0 \\ -0.3867 \\ 0.8934 \\ -0.2282 \end{pmatrix}$$

$$\hat{u} \cdot \widehat{u^T} = \begin{pmatrix} 0 & 0 & 0 & 0 \\ 0 & 0.1495 & -0.3455 & 0.0882 \\ 0 & -0.3455 & 0.7983 & -0.2039 \\ 0 & 0.0882 & -0.2039 & 0.0520 \end{pmatrix}$$

$$I - 2 \cdot \hat{u} \cdot \widehat{u^T} = \begin{pmatrix} 1 & 0 & 0 & 0 \\ 0 & 0.7008 & 0.6911 & -0.1765 \\ 0 & 0.6911 & -0.5966 & 0.4078 \\ 0 & -0.1765 & 0.4078 & 0.8958 \end{pmatrix}$$

Let us name this other Householder transformation matrix $\boldsymbol{Q_2}$. Now let us find out what happens if we multiply this matrix by the one we had before:

$$Q_2 \cdot (Q_1 \cdot A) = \begin{pmatrix} 1 & 0 & 0 & 0 \\ 0 & 0.7008 & 0.6911 & -0.1765 \\ 0 & 0.6911 & -0.5966 & 0.4078 \\ 0 & -0.1765 & 0.4078 & 0.8958 \end{pmatrix} \cdot \begin{pmatrix} 11.489 & 10.792 & 10.879 & 4.526 \\ 0 & 2.1617 & 3.2487 & 5.9498 \\ 0 & 2.1319 & -1.2587 & 5.6489 \\ 0 & -0.5446 & 3.2387 & -0.4513 \end{pmatrix}$$

$$Q_2 \cdot (Q_1 \cdot A) = \begin{pmatrix} 11.489 & 10.792 & 10.879 & 4.526 \\ 0 & 3.0846 & 0.8349 & 8.1536 \\ 0 & 0 & 4.3173 & 0.5579 \\ 0 & 0 & 1.8144 & 0.8491 \end{pmatrix}$$

which is another step into making our matrix A into a triangular matrix. Each Householder transformation takes us one step forward. The idea is to make several reflections such that the result will be an upper triangular matrix, which is matrix R. If we will continue doing this procedures with the last matrix we obtained we will get:

$$R = Q_3 \cdot Q_2 \cdot Q_1 \cdot A$$

You can see that we are multiplying three orthonormal matrices times A to get R. However, these three orthonormal matrices can be compacted into another matrix that will also be orthonormal. The proof of this is shown next:

If A and B are two orthonormal matrices, then their multiplication is also orthonormal:

$$A^T \cdot A = A \cdot A^T = IB^T \cdot B = B \cdot B^T = I$$

$$(A \cdot B) \cdot (A \cdot B)^T = (A \cdot B) \cdot (B^T \cdot A^T) = A \cdot (B \cdot B^T) \cdot A = A \cdot I \cdot A^T = A \cdot A^T = I$$

From here we can say that these three matrices can be compacted into a single matrix which will be the transpose of a matrix of name Q, to get:

$$R = Q^T \cdot A \rightarrow Q^T = Q_3 \cdot Q_2 \cdot Q_1$$

And if would multiply by matrix Q on both sides, we would get the QR decomposition of matrix A:

$$A = Q \cdot R$$

To finish this problem, we will show the results:

$$Q_3 = \begin{pmatrix} 1 & 0 & 0 & 0 \\ 0 & 1 & 0 & 0 \\ 0 & 0 & 0.9218 & 0.3874 \\ 0 & 0 & 0.3874 & -0.9218 \end{pmatrix}$$

$$R = Q_3 \cdot Q_2 \cdot Q_1 \cdot A = \begin{pmatrix} 11.489 & 10.792 & 10.879 & 4.526 \\ 0 & 3.0846 & 0.8349 & 8.1536 \\ 0 & 0 & 4.6831 & 0.8433 \\ 0 & 0 & 0 & -0.5666 \end{pmatrix}$$

$$Q^T = Q_3 \cdot Q_2 \cdot Q_1 = \begin{pmatrix} 0.0870 & 0.0870 & 0.6092 & 0.7833 \\ 0.3437 & 0.6680 & 0.4617 & -0.4715 \\ 0.3770 & 0.5328 & -0.6436 & 0.3995 \\ -0.8555 & 0.5121 & -0.0361 & 0.0662 \end{pmatrix}$$

This is the Householder method to solve for the QR decomposition. However, another method still exists which is the ***Givens rotation***. Givens rotation is a method in which the column vectors of a matrix are rotated consecutively in such a way that the result is an upper triangular matrix. This method will not be explained further since it is not used extensively nowadays. It has very detailed and closed applications. It was mentioned however, as you might be interested in researching on your own. We also mentioned this because each of them has its advantages and applications. For example, Gram Schmidt is mostly used in small systems that do not require much accuracy on the results, because it is numerically unstable for computers since its process gives rise to computational errors. On the other hand, Householder method is much more stable and is well applicable to greater systems; however, it has the slight disadvantage that it is not parallelizable. Just the same, Householder reflections is an amazing algorithm to compute QR decomposition.

As a recap, in this chapter we saw we can decompose matrices in many forms, depending on our needs and wants. In this chapter we saw four ways of decomposing matrices. However, many methods still exist, which we will study in other chapters, since we need to master other concepts before that. The most famous today is the Singular Value Decomposition, although each method has its own advantages. One of the concepts we must master before proceeding are ***determinants***.

DETERMINANTS

Determinants are a very important number that has many information packed within it. However, determinants also has a certain of mysterious origin. Some authors say that determinants came before matrices. Others say that determinants are related to permutations. In our case we saw that determinants represent the volume distortion of an n-dimensional cube during a transformation. But why? Why is the determinant explained that way? Why a system of equations with zero determinant do not have full rank? How did people find such a number? How did people discover how to compute for a determinant?

Many of these questions come to mind. To answer these questions we need to follow up the determinant and look where it appears, we need to study the history of determinants. Determinants appear in many places, and it is true that they represent the distortion of a volume and they are related with permutations, however, determinants have their origins in systems of equations. From previous experiences, you probably know already that the determinant is a number that correlates all elements in a matrix in a certain way, and why determinants do this is what we are about to discuss.

Determinants were first noticed while obtaining analytical solutions to systems of equations. One of the examples we saw before is the 2x2 system:

$$\begin{pmatrix} a_1 & a_2 \\ a_3 & a_4 \end{pmatrix}\begin{pmatrix} x \\ y \end{pmatrix} = \begin{pmatrix} b_1 \\ b_2 \end{pmatrix}$$

And its solution is:

$$x = \frac{a_4 b_1 - a_2 b_2}{a_1 \cdot a_4 - a_2 \cdot a_3} \quad y = \frac{-a_3 b_1 + a_1 b_2}{a_1 \cdot a_4 - a_2 \cdot a_3}$$

Where the common denominator in our variables, that appears when solving systems of equations analytically, is the determinant. However, computing this solutions is not easy for systems of high order. For example, if we would have a 3x3 system of equations of the following form (assuming there are nonzero entries across the main diagonal):

$$\begin{pmatrix} a_{11} & a_{12} & a_{13} \\ a_{21} & a_{22} & a_{23} \\ a_{31} & a_{32} & a_{33} \end{pmatrix}\begin{pmatrix} x \\ y \\ z \end{pmatrix} = \begin{pmatrix} b_1 \\ b_2 \\ b_3 \end{pmatrix}$$

Then the solution would be the following:

$$x = \frac{(a_{22} \cdot a_{33} - a_{32} \cdot a_{23}) b_1 - (a_{12} \cdot a_{33} - a_{13} \cdot a_{32}) b_2 + (a_{12} \cdot a_{23} - a_{13} \cdot a_{22}) b_3}{a_{11} \cdot a_{22} \cdot a_{33} + a_{12} \cdot a_{23} \cdot a_{31} + a_{13} \cdot a_{21} \cdot a_{32} - a_{31} \cdot a_{22} \cdot a_{13} - a_{32} \cdot a_{23} \cdot a_{13} - a_{33} \cdot a_{21} \cdot a_{12}}$$

$$y = \frac{(a_{23} \cdot a_{31} - a_{21} \cdot a_{33}) b_1 - (a_{11} \cdot a_{33} - a_{13} \cdot a_{31}) b_2 + (a_{21} \cdot a_{13} - a_{11} \cdot a_{23}) b_3}{a_{11} \cdot a_{22} \cdot a_{33} + a_{12} \cdot a_{23} \cdot a_{31} + a_{13} \cdot a_{21} \cdot a_{32} - a_{31} \cdot a_{22} \cdot a_{13} - a_{32} \cdot a_{23} \cdot a_{13} - a_{33} \cdot a_{21} \cdot a_{12}}$$

And the determinant as we can see is a lot bigger than for a 2x2 system. But let us talk more about the determinant itself, the fact that the determinant appears in the denominator is key, because the determinant only causes trouble if it is equated to zero. So when the determinant is equated to zero, something special must happen. And indeed it does, because when the ***determinant is equal to zero*** then we see that we ***do not have a solution*** to the system because we would be dividing by zero while calculating our variables.

Notice that the determinant of a matrix only depends on the matrix coefficients, and therefore, gives only information about the matrix. This allows us to see how a system will behave in case it has a matrix with zero determinant.

However, from what we have seen previously, we also know that in a system of equations we will not have a single solution if we have independent columns or rows among the coefficient matrix. Therefore, we can correlate these two concepts.

From this point of view, we can now view the determinant as a number that indicates whether there is or no linear dependency among the rows or columns of a square matrix, as well as pointing out if there is or no single solution to a system of equations.

If the reader has tried to solve linear systems of equations analytically, he will probably think that it is not an easy task. However, this is not the only way to compute an expression for determinants of a matrix. There exists Laplace method which splits a determinant of a $n \times n$ matrix into ***n*** determinants of $(n-1)\times(n-1)$ matrices. This represents a lot of computing time, since to calculate a determinant of a $n \times n$ matrix by Laplace method, we would need to compute determinants to get the answer, besides all sums and multiplications that need to be done. For this, we rely on some properties of determinants to simplify.

PROPERTIES OF DETERMINANTS

Even though determinants are somewhat strange to compute, there are many properties that can help us along the way when computing them. But in order to deduce them we will need to rely on four main properties. By the way, to denote the determinant of a matrix ***A***, we will denote it with two vertical bars at the sides, like this $|A|$. Having this said, let us introduce the properties:

- ***Property 1:*** The determinant of the Identity matrix is 1. Therefore, $|I|=1$

- ***Property 2:*** If we exchange two rows from a matrix, then the determinant changes sign. Therefore,

$$A = \begin{pmatrix} a_{11} & a_{12} & \cdots & a_{1m} \\ a_{21} & a_{22} & \cdots & a_{2m} \\ \vdots & \vdots & \cdots & \vdots \\ a_{r1} & a_{r2} & \cdots & a_{rm} \\ a_{t1} & a_{t2} & \cdots & a_{tm} \\ \vdots & \vdots & \cdots & \vdots \\ a_{n1} & a_{n2} & \cdots & a_{nm} \end{pmatrix}. Exchanging the r^{th} row wiht the t^{th} row we get:$$

$$B = \begin{pmatrix} a_{11} & a_{12} & \cdots & a_{1m} \\ a_{21} & a_{22} & \cdots & a_{2m} \\ \vdots & \vdots & \cdots & \vdots \\ a_{t1} & a_{t2} & \cdots & a_{tm} \\ a_{r1} & a_{r2} & \cdots & a_{rm} \\ \vdots & \vdots & \cdots & \vdots \\ a_{n1} & a_{n2} & \cdots & a_{nm} \end{pmatrix}. then |B| = -|A|$$

- ***Property 3:*** If we have a constant ***c*** multiplying a row of a matrix, then we could extract it and compute the determinant without the constant, therefore,

$$A = \begin{pmatrix} a_{11} & a_{12} & \cdots & a_{1m} \\ a_{21} & a_{22} & \cdots & a_{2m} \\ \vdots & \vdots & \cdots & \vdots \\ a_{n1} & a_{n2} & \cdots & a_{nm} \end{pmatrix}; B = \begin{pmatrix} a_{11} & a_{12} & \cdots & a_{1m} \\ c\cdot a_{21} & c\cdot a_{22} & \cdots & c\cdot a_{2m} \\ \vdots & \vdots & \cdots & \vdots \\ a_{n1} & a_{n2} & \cdots & a_{nm} \end{pmatrix}.$$

$$|B| = c\cdot|A|$$

- ***Property 4:*** If we have a determinant of the form shown next, then we are able to separate it like this:

$$\begin{vmatrix} a+a' & b+b' \\ c & d \end{vmatrix} = \begin{vmatrix} a & b \\ c & d \end{vmatrix} + \begin{vmatrix} a' & b' \\ c & d \end{vmatrix}$$

However, this does not mean that the determinant is a linear function since:

$$|(A+B)| \neq |A| + |B|$$

- ***Property 5:*** If two rows are equal in a matrix, then its determinant is zero. This can be deduced from property 2. Suppose a matrix has 2 equal rows and we exchange them. By doing this the determinant changes sign. But since we have the same matrix as when we started therefore, we should have the same determinant. The only way for this to be true is if the determinant is 0. Therefore, ***if a matrix has two equal rows, then the determinant is zero***.

- ***Property 6:*** If we add or subtract a multiple ***c*** of a row to another row, then the determinant does not change. This can be shown from property 3 and 4, because:

$$A = \begin{pmatrix} a_{11} & a_{12} & \cdots & a_{1m} \\ a_{21} & a_{22} & \cdots & a_{2m} \\ \vdots & \vdots & \cdots & \vdots \\ a_{r1} & a_{r2} & \cdots & a_{rm} \\ a_{t1} & a_{t2} & \cdots & a_{tm} \\ \vdots & \vdots & \cdots & \vdots \\ a_{n1} & a_{n2} & \cdots & a_{nm} \end{pmatrix} \to R_t = R_t - c \cdot R_r B$$

$$= \begin{pmatrix} a_{11} & a_{12} & \cdots & a_{1m} \\ a_{21} & a_{22} & \cdots & a_{2m} \\ \vdots & \vdots & \cdots & \vdots \\ a_{r1} & a_{r2} & \cdots & a_{rm} \\ a_{t1} - c \cdot a_{r1} & a_{t2} - c \cdot a_{r2} & \cdots & a_{tm} - c \cdot a_{rm} \\ \vdots & \vdots & \cdots & \vdots \\ a_{n1} & a_{n2} & \cdots & a_{nm} \end{pmatrix}$$

$$det(B) = \begin{vmatrix} a_{11} & a_{12} & \cdots & a_{1m} \\ a_{21} & a_{22} & \cdots & a_{2m} \\ \vdots & \vdots & \cdots & \vdots \\ a_{r1} & a_{r2} & \cdots & a_{rm} \\ a_{t1} - c \cdot a_{r1} & a_{t2} - c \cdot a_{r2} & \cdots & a_{tm} - c \cdot a_{rm} \\ \vdots & \vdots & \cdots & \vdots \\ a_{n1} & a_{n2} & \cdots & a_{nm} \end{vmatrix} = \begin{vmatrix} a_{11} & a_{12} & \cdots & a_{1m} \\ a_{21} & a_{22} & \cdots & a_{2m} \\ \vdots & \vdots & \cdots & \vdots \\ a_{r1} & a_{r2} & \cdots & a_{rm} \\ a_{t1} & a_{t2} & \cdots & a_{tm} \\ \vdots & \vdots & \cdots & \vdots \\ a_{n1} & a_{n2} & \cdots & a_{nm} \end{vmatrix} - c \begin{vmatrix} a_{11} & a_{12} & \cdots & a_{1} \\ a_{21} & a_{22} & \cdots & a_{2} \\ \vdots & \vdots & \cdots & \vdots \\ a_{r1} & a_{r2} & \cdots & a_{r} \\ a_{r1} & a_{r2} & \cdots & a_{r} \\ \vdots & \vdots & \cdots & \vdots \\ a_{n1} & a_{n2} & \cdots & a_{n} \end{vmatrix}$$

The last operation comes from the fourth property which allows us to separate the determinant into two determinants. But we can see that the second determinant has a repeated row, therefore, its determinant is zero. From this we can conclude that:

$$det(B) = det(A)$$

This property allows us to make linear combination of the rows, and reduce a matrix without changing the determinant.

- ***Property 7:*** If a matrix has a row of zeros then the determinant of the matrix must be zero. This property can be shown by multiple ways from the preceding ones.

$$A=\begin{pmatrix} a_{11} & a_{12} & \cdots & a_{1m} \\ a_{21} & a_{22} & \cdots & a_{2m} \\ \vdots & \vdots & \cdots & \vdots \\ 0 & 0 & \cdots & 0 \\ \vdots & \vdots & \cdots & \vdots \\ a_{n1} & a_{n2} & \cdots & a_{nm} \end{pmatrix} \rightarrow |A| = 0.$$

- ***Property 8:*** If a matrix is an upper or lower triangular matrix, then the determinant of the matrix is the product of the coefficients in the main diagonal. This can be shown from the sixth, third and first property. Since if we have a triangular matrix we can do row operations and turn the matrix into a diagonal matrix without changing the determinant. From the third property we can factor out all the constants in the diagonal, and we would have the identity matrix. Therefore,

$$|A| = |\begin{matrix} a_{11} & a_{12} & \cdots & a_{1m-1} & a_{1m} \\ 0 & a_{22} & \cdots & a_{2m-1} & a_{2m} \\ \vdots & \vdots & \cdots & \cdots & \vdots \\ 0 & 0 & \cdots & 0 & a_{nm} \end{matrix} |applyingrowoperations:$$

$$|A| = |\begin{matrix} a_{11} & 0 & \cdots & 0 & 0 \\ 0 & a_{22} & \cdots & 0 & 0 \\ \vdots & \vdots & \cdots & \cdots & \vdots \\ 0 & 0 & \cdots & 0 & a_{nm} \end{matrix} |Factoringouttheconstants:$$

$$|A| = a_{11} \cdot a_{22} \cdot \ldots \cdot a_{nn} \cdot |\begin{matrix} 1 & 0 & \cdots & 0 & 0 \\ 0 & 1 & \cdots & 0 & 0 \\ \vdots & \vdots & \cdots & \cdots & \vdots \\ 0 & 0 & \cdots & 0 & 1 \end{matrix}| = a_{11} \cdot a_{22} \cdot \ldots \cdot a_{nn}$$

The same procedure applies if ***A*** is a lower triangular matrix.

- ***Property 9:*** We have seen this property in chapter 1. The determinant of the product of matrices is the product of the determinants of matrices:

$$|A \cdot B| = |A| \cdot |B|$$

- ***Property 10:*** The determinant of a matrix is equal to the determinant of its transpose. This property is very simple but it allows us to work with determinants better.

$$|A^T| = |A|$$

This property can be shown assuming the A has an LU factorization. Thus,

$$\left|(L \cdot U)^T\right| = \left|L \cdot U\right| \left|U^T \cdot L^T\right| = \left|L \cdot U\right|$$

Separating both determinants, we get:

$$\left|U^T\right| \cdot \left|L^T\right| = \left|L\right| \cdot \left|U\right|$$

And as L and U are triangular matrices then their determinants are the product of the elements in their main diagonals. Transposing a matrix does not alter the main diagonal, therefore, both determinants are the same and the proof is complete.

This property may sound simple, but it allows as to work with row operations and with column operation, since transposing a matrix does not change the determinant.

CHAPTER 5

EIGENVALUES AND EIGENVECTORS

"Eigenvalues are in everything. There's an eigenvalue in the burrito you're going to eat for lunch today"

mathprofessorquotes.com

INTRODUCTION

From past chapters, we have learned relevant knowledge on many problems and applications of algebra. Algebra appears in many places, from systems of linear differential equations to engineering problems, numerical simulations and more. However, we still have not discussed how to approach these problems. For this, we need to learn first about a special set of vectors called the eigenvectors which are correlated with the eigenvalues.

Eigenvalues and eigenvectors may sound like the solution to any problem, but that is not the case. They are just tools which serve very well to solve most of the problems we have just mentioned now, and many of their properties correlate with the knowledge we now possess. By the way, the word eigenvalue comes from both German and English, "eigen" means "own," and "value" is a "quantity." Therefore, eigenvalues are "own quantities." We will discuss this term in a little while.

THE MAIN PROBLEM

Eigenvalues and eigenvectors are a special set of values and vectors that appear in many problems and situations that go from abstract algebra to real world applications like engineering. As we have said before, the word eigenvalue means "own quantity," but who or what owns an eigenvalue? To answer this question, think of a simple idea that all matrices (with full rank) are different since any change in a coefficient or size would change the matrix itself. From here we can answer the question since any matrix has its own eigenvalues that will depend on the matrix itself.

To see where an eigenvalue comes from, let us look at the following. Suppose we have a matrix transformation ***A*** that takes a vector ***v*** and gives another vector ***u***, therefore,

$$A\vec{v} = \vec{u}$$

However, let us also suppose that we are interested in all the vectors whose transformation have the same direction as their input, vectors that do not necessarily need to have the same length nor orientation, but yes, the same direction. Therefore, a vector ***u*** that satisfies this condition is the following:

$$\vec{u} = \ddot{e}\vec{v}$$

We can see that vectors ***u*** and ***v*** have the same direction, but there is a factor ***lambda*** that will scale vector ***v*** to become vector ***u***. Using our past equation we can say that

$$A\vec{v} = \ddot{e}\vec{v}$$

THIS CAN BE REARRANGED IN THE FOLLOWING WAY:

$$A\vec{v} - \lambda\vec{v} = \vec{0}$$

$$(A - \lambda I)\vec{v} = \vec{0}$$

The above expression is the beginning of eigenvalues and eigenvectors, and you may already be familiar with it. However, let us take a closer look at it. We can obviously see that we have a new transformation $(A - \ddot{e}I)$

that takes a vector ***v*** and transforms it into the zero vector. Therefore, we could interpret that for any vector ***v***, different from zero, ***v*** will belong in the Null space of $(A-\text{ë}I)$, therefore, $(A-\text{ë}I)$ is not a full rank matrix , and therefore, it has linearly dependent rows and its determinant must be zero.

Remember that our main goal here is to find a vector who's transformation has the same direction as its input. However, to do this we must first find the term lambda, since the vector we are after, vector ***v***, belongs in the null space of $(A-\text{ë}I)$. Therefore, we need to compute a lambda such that $(A-\text{ë}I)$ is not full rank. For this, we will rely on the interpretation we did before, which is "the determinant of $(A-\text{ë}I)$ must be zero:"

$$det(A-\text{ë}I)=0$$

From the above equation we can compute lambda, then input that value into our original matrix $(A-\text{ë}I)$ and calculate its null space, thereby obtaining vector ***v***. From these conditions, we name all lambdas that satisfy the above expression as ***eigenvalues,*** and every vector ***v*** is called an ***eigenvector***. We made special mention to the word eigenvalue since, from what we can see now, every matrix will have its own eigenvalues and will depend only on the matrix. To explain this graphically, let us look at the following example. Suppose we have the following transformation ***A*** and vector ***v***:

$$A=\begin{pmatrix} 2 & 0 \\ 0 & 4 \end{pmatrix}; \vec{v}=\begin{pmatrix} 5 \\ 2 \end{pmatrix}$$

The transformation of vector ***v***, let us call it vector ***u,*** is

$$\vec{u}=\begin{pmatrix} 10 \\ 8 \end{pmatrix}$$

Graphically, vector ***v*** (colored in green) and its transformation (colored in red) would look like this:

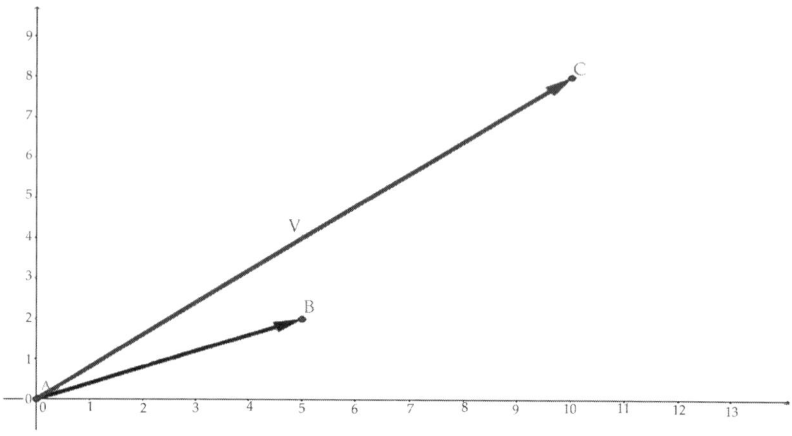

Figure 15: Vector transformation.

We can see that vector $\boldsymbol{v}$ and vector $\boldsymbol{u}$ do not have the same direction, and therefore, $\boldsymbol{v}$ is not an eigenvector of $\boldsymbol{A}$. However, if we have another vector, let us call it $\boldsymbol{a}$, and its transformation, let us call it vector $\boldsymbol{b}$, of the following form:

$$\vec{a} = \begin{pmatrix} 2 \\ 0 \end{pmatrix}; \vec{b} = A\vec{a} = \begin{pmatrix} 2 & 0 \\ 0 & 4 \end{pmatrix} \begin{pmatrix} 2 \\ 0 \end{pmatrix} = \begin{pmatrix} 4 \\ 0 \end{pmatrix}$$

then both have the same direction, because vector $\boldsymbol{b}$ is 2 times vector $\boldsymbol{a}$; therefore, vector $\boldsymbol{a}$ is an eigenvector of matrix $\boldsymbol{A}$. That was a very simple example. Let us do something more complex to show how to compute eigenvectors and eigenvalues. Suppose we have the following matrix $\boldsymbol{B}$ and we want to calculate its eigenvalues and eigenvectors:

$$B = \begin{pmatrix} 1 & 0 & 4 \\ 0 & 2 & 0 \\ 3 & 1 & -3 \end{pmatrix}$$

To compute the eigenvalues of matrix $\boldsymbol{B}$ we must do the following determinant:

$$det(B-\lambda I)=\begin{vmatrix}1-\lambda & 0 & 4 \\ 0 & 2-\lambda & 0 \\ 3 & 1 & -3-\lambda\end{vmatrix}=0$$

Therefore,

$$det(B-\lambda I)=(1-\lambda)\cdot(2-\lambda)\cdot(-3-\lambda)-3\cdot 4\cdot(2-\lambda)$$

$$det(B-\lambda I)=(\lambda-2)\cdot(-\lambda^2-2\lambda+15)=0$$

$$det(B-\lambda I)=(\lambda-2)\cdot(\lambda-3)\cdot(\lambda+5)=0$$

$$\lambda=2,3,-5$$

We can see that we have three different eigenvalues, and all satisfy the above condition. By the way, the polynomial made by the above determinant is called the characteristic polynomial and its degree will always be equal to the size of the matrix; therefore, an $n\times n$ matrix will always have an ***n*** degree characteristic polynomial. From the fundamental theorem of algebra, we know that an ***n*** degree polynomial will have ***n*** roots, whether they are real or complex, all different or repeated; therefore, an $n\times n$ matrix will have ***n*** eigenvalues.

After calculating the different values of lambda, we must introduce them one at a time into our original matrix and calculate its null space, and therefore, the eigenvectors. As a custom for this book we will name the eigenvectors $\vec{e_i}$ depending on the number of eigenvectors. Doing this with matrix B, we get:

$$(B-\lambda_1 I)\vec{e_1}=\vec{0}\rightarrow\begin{pmatrix}1-2 & 0 & 4 \\ 0 & 2-2 & 0 \\ 3 & 1 & -3-2\end{pmatrix}\begin{pmatrix}e_{11} \\ e_{12} \\ e_{13}\end{pmatrix}=\begin{pmatrix}0 \\ 0 \\ 0\end{pmatrix}$$

$$\begin{pmatrix} -1 & 0 & 4 \\ 0 & 0 & 0 \\ 3 & 1 & -5 \end{pmatrix}\begin{pmatrix} e_{11} \\ e_{12} \\ e_{13} \end{pmatrix} = \begin{pmatrix} 0 \\ 0 \\ 0 \end{pmatrix} \rightarrow \textit{Applying row reduction}$$

$$\begin{pmatrix} -1 & 0 & 4 \\ 0 & 0 & 0 \\ 0 & 1 & 7 \end{pmatrix}\begin{pmatrix} e_{11} \\ e_{12} \\ e_{13} \end{pmatrix} = \begin{pmatrix} 0 \\ 0 \\ 0 \end{pmatrix}$$

$$e_{11} = 4e_{13}$$
$$e_{12} = -7e_{13}$$

Therefore, vectoris of the following form:

$$\begin{pmatrix} e_{11} \\ e_{12} \\ e_{13} \end{pmatrix} = \begin{pmatrix} 4 \cdot e_{13} \\ -7 \cdot e_{13} \\ e_{13} \end{pmatrix} = e_{13} \cdot \begin{pmatrix} 4 \\ -7 \\ 1 \end{pmatrix}$$

Doing this same procedure with the other 2 eigenvalues, we get

$$(B - \lambda_2 I)\vec{e_2} = \vec{0} = \begin{pmatrix} 1-3 & 0 & 4 \\ 0 & 2-3 & 0 \\ 3 & 1 & -3-3 \end{pmatrix}\begin{pmatrix} e_{21} \\ e_{22} \\ e_{23} \end{pmatrix} = \begin{pmatrix} 0 \\ 0 \\ 0 \end{pmatrix}$$

$$\begin{pmatrix} -2 & 0 & 4 \\ 0 & -1 & 0 \\ 3 & 1 & -6 \end{pmatrix}\begin{pmatrix} e_{21} \\ e_{22} \\ e_{23} \end{pmatrix} = \begin{pmatrix} 0 \\ 0 \\ 0 \end{pmatrix} \rightarrow \textit{Applying row reduction}$$

$$\begin{pmatrix} -2 & 0 & 4 \\ 0 & -1 & 0 \\ 0 & 0 & 0 \end{pmatrix}\begin{pmatrix} e_{21} \\ e_{22} \\ e_{23} \end{pmatrix} = \begin{pmatrix} 0 \\ 0 \\ 0 \end{pmatrix}$$

$$e_{21} = 2e_{23}$$
$$e_{22} = 0$$
$$e_{23} = e_{23}$$

Therefore, vector $\overrightarrow{e_2}$ is of the following form:

$$\begin{pmatrix} e_{21} \\ e_{22} \\ e_{23} \end{pmatrix} = \begin{pmatrix} 2 \cdot e_{23} \\ 0 \\ e_{23} \end{pmatrix} = e_{23} \cdot \begin{pmatrix} 2 \\ 0 \\ 1 \end{pmatrix}$$

$$(B - \lambda_2 I)\vec{e_2} = \vec{0} = \begin{pmatrix} 1-3 & 0 & 4 \\ 0 & 2-3 & 0 \\ 3 & 1 & -3-3 \end{pmatrix} \begin{pmatrix} e_{21} \\ e_{22} \\ e_{23} \end{pmatrix} = \begin{pmatrix} 0 \\ 0 \\ 0 \end{pmatrix}$$

$$\begin{pmatrix} 6 & 0 & 4 \\ 0 & 7 & 0 \\ 3 & 1 & 2 \end{pmatrix} \begin{pmatrix} e_{31} \\ e_{32} \\ e_{33} \end{pmatrix} = \begin{pmatrix} 0 \\ 0 \\ 0 \end{pmatrix} \rightarrow Applying\ row\ reduction$$

$$\begin{pmatrix} 6 & 0 & 4 \\ 0 & 1 & 0 \\ 0 & 0 & 0 \end{pmatrix} \begin{pmatrix} e_{31} \\ e_{32} \\ e_{33} \end{pmatrix} = \begin{pmatrix} 0 \\ 0 \\ 0 \end{pmatrix}$$

$$e_{31} = \frac{-2}{3} e_{33}$$

$$e_{32} = 0$$

$$e_{33} = e_{33}$$

Therefore, vector $\vec{e_3}$ is of the following form:

$$\begin{pmatrix} e_{31} \\ e_{32} \\ e_{33} \end{pmatrix} = \begin{pmatrix} -2/3 \cdot e_{33} \\ 0 \\ e_{33} \end{pmatrix} = e_{33} \cdot \begin{pmatrix} -2/3 \\ 0 \\ 1 \end{pmatrix}$$

And from here we can see that our eigenvectors are:

$$\vec{e_1} = \begin{pmatrix} 4 \\ -7 \\ 1 \end{pmatrix}; \vec{e_2} = \begin{pmatrix} 2 \\ 0 \\ 1 \end{pmatrix}; e_3 = \begin{pmatrix} -2/3 \\ 0 \\ 1 \end{pmatrix}$$

A detail here is that these are the eigenvectors of ***B***. However, any multiple of them will also be an eigenvector. So when we say we found the eigenvectors, in reality we are saying that we found a basis for them. For example, the following vector is also an eigenvector:

$$2\cdot\vec{e_2} = \begin{pmatrix} 4 \\ 0 \\ 2 \end{pmatrix}$$

But it is better to keep on with the ones calculated before. In some occasions, because of this conjecture, some authors decide to normalize their eigenvectors. By the way, a scalar times an eigenvector will still be an eigenvector, but a linear combination of the eigenvectors will not necessarily be an eigenvector. Now let us look at something interesting that we already know but is probably not realized yet. What happens if we multiply ***B*** by an eigenvector? Let us check it out:

$$B\cdot\vec{e_1} = \begin{pmatrix} 1 & 0 & 4 \\ 0 & 2 & 0 \\ 3 & 1 & -3 \end{pmatrix}\cdot\begin{pmatrix} 4 \\ -7 \\ 1 \end{pmatrix} = \begin{pmatrix} 8 \\ -14 \\ 2 \end{pmatrix} = 2\cdot\begin{pmatrix} 4 \\ -7 \\ 1 \end{pmatrix} = \lambda_1\cdot\vec{e_1}$$

which is exactly our formulation for eigenvectors and eigenvalue. If we have a matrix together with its eigenvalues and its eigenvectors, then if we multiply the matrix by an eigenvector the result will be the corresponding eigenvalue times the eigenvector! Therefore,

If we have a matrix ***A*** with eigenvalues $\boldsymbol{\lambda_1, \lambda_2, \ldots, \lambda_i}$, and eigenvectors $\boldsymbol{e_1, e_2, \ldots, e_i}$

$$A\cdot\vec{e_i} = \lambda_i\vec{e_i}$$

As we said before, a characteristic polynomial is a polynomial of degree ***n*** which has ***n*** roots that could be either real or complex. If the roots are real then everything makes sense -- a vector is transformed and his transformation is the same vector multiplied by a constant. However, what happens when an eigenvalue is imaginary or complex? Let us see this question with a quick example. Suppose we have the following matrix ***C*** and we want to calculate its eigenvalues:

$$C = \begin{pmatrix} 0 & 1 \\ -1 & 0 \end{pmatrix}$$

C is a matrix that rotates every vector in the plane 90° degrees counterclockwise (hint: this is one of the rotation matrices we mentioned on a previous chapter), so in this case, how can this matrix transform a vector such that its result will have the same direction? Is it possible or impossible? Actually, it is possible! Let us calculate the eigenvectors of this matrix.

$$det(C-\lambda I)=\begin{vmatrix}0-\lambda & 1\\ -1 & 0-\lambda\end{vmatrix}=\begin{vmatrix}-\lambda & 1\\ -1 & -\lambda\end{vmatrix}=0$$

Solving the determinant we get that:

$$det(C-\ddot{e}I)=\ddot{e}^2+1=0$$

Therefore, the lambdas are:

$$\lambda=\sqrt{-1}$$

$$\lambda_1=i;\lambda_2=-i$$

Things are getting a little strange now, but still, let us calculate the eigenvectors:

$$(C-\lambda_1 I)\vec{d}=\begin{pmatrix}0-i & 1\\ -1 & 0-i\end{pmatrix}\begin{pmatrix}d_1\\ d_2\end{pmatrix}=\begin{pmatrix}-i & 1\\ -1 & -i\end{pmatrix}\begin{pmatrix}d_1\\ d_2\end{pmatrix}=0$$

If we do row reduction multiplying the first row times ***i*** and add it to the second row, we get:

$$\begin{pmatrix}-i & 1\\ 0 & 0\end{pmatrix}\begin{pmatrix}d_1\\ d_2\end{pmatrix}=0$$

$$d_2=i\cdot d_1$$

$$\vec{d}=\begin{pmatrix}d_1\\ i\cdot d_1\end{pmatrix}=d_1\cdot\begin{pmatrix}1\\ i\end{pmatrix}$$

Doing the same procedure to find the other eigenvector, we get that both eigenvectors have the following form:

$$\vec{e_1} = \begin{pmatrix} 1 \\ i \end{pmatrix}; \vec{e_2} = \begin{pmatrix} i \\ 1 \end{pmatrix}$$

It is true that we cannot graph complex vectors as we usually do with real vectors. However, we see that all of these procedures still endure even when this imaginary things happen, because those are true eigenvectors of matrix ***C***. From this example we can see that imaginary eigenvalues are somehow related to vector rotations, but this is just a quick example to check that eigenvalues can also be complex.

There is also a very important, and sometimes disastrous, phenomenon that can occur while calculating eigenvalues and eigenvectors. This is not a catastrophe but it limits the possibilities of working with a matrix and get information from it.

For example, if we have a matrix of $n \times n$ size, it will have ***n*** eigenvalues. Supposing that the ***n*** eigenvalues are different, then that assures that we will have ***n*** different eigenvectors. But if we have repeated eigenvalues, then there exists the possibility that we do not have ***n*** eigenvalues. To show this, two examples are shown.

Suppose we have the following matrix ***D***, and we want to calculate its eigenvectors:

$$D = \begin{pmatrix} 2 & 1 \\ 0 & 2 \end{pmatrix}$$

We do the determinant of $(D - \lambda I)$ and get:

$$det(D - \lambda I) = (2 - \lambda)^2$$

$$\lambda_1 = 2; \lambda_2 = 2$$

Where we already see that we have a repeated eigenvalue. Let us calculate the eigenvectors:

$$(D-\lambda_1 I)\vec{d}=\begin{pmatrix}2-2 & 1\\ 0 & 2-2\end{pmatrix}\begin{pmatrix}d_1\\ d_2\end{pmatrix}=\begin{pmatrix}0\\ 0\end{pmatrix}$$

$$\vec{e_1}=d_1\cdot\begin{pmatrix}1\\ 0\end{pmatrix}$$

If we input λ_2 we will get the same eigenvector, therefore, we have two same eigenvalues and one linearly independent eigenvector. This is not a desirable situation. However, this does not always happen, like in the following example.

Suppose we have the following matrix ***E***, and we want to calculate its eigenvectors:

$$E=\begin{pmatrix}1 & -1 & -1\\ -1 & 1 & -1\\ -1 & -1 & 1\end{pmatrix}$$

Therefore, doing the same procedure we get that: where is/are the equation/s?

If we input the first eigenvalue to obtain an eigenvector, we get

$$(E-\lambda_1 I)\cdot\vec{a}=\begin{pmatrix}2 & -1 & -1\\ -1 & 2 & -1\\ -1 & -1 & 2\end{pmatrix}\begin{pmatrix}a_1\\ a_2\\ a_3\end{pmatrix}=\vec{0}\rightarrow\vec{e_1}=\begin{pmatrix}1\\ 1\\ 1\end{pmatrix}$$

Introducing the other eigenvalue, we get

$$(E-\lambda_2 I)\cdot\vec{b}=\begin{pmatrix}-1 & -1 & -1\\ -1 & -1 & -1\\ -1 & -1 & -1\end{pmatrix}\cdot\begin{pmatrix}b_1\\ b_2\\ b_3\end{pmatrix}=\vec{0}$$

And if we do row reduction to calculate the eigenvector, we get

$$\begin{pmatrix} -1 & -1 & -1 \\ 0 & 0 & 0 \\ 0 & 0 & 0 \end{pmatrix} \cdot \begin{pmatrix} b_1 \\ b_2 \\ b_3 \end{pmatrix} = \vec{0}$$

$$b_1 = -b_2 - b_3$$

$$\vec{b} = \begin{pmatrix} b_1 \\ b_2 \\ b_3 \end{pmatrix} = \begin{pmatrix} -b_2 - b_3 \\ b_2 \\ b_3 \end{pmatrix} = b_2 \cdot \begin{pmatrix} -1 \\ 1 \\ 0 \end{pmatrix} + b_3 \cdot \begin{pmatrix} -1 \\ 0 \\ 1 \end{pmatrix}$$

As you can see, we were able to get three linearly independent eigenvectors which are:

$$\vec{e_1} = \begin{pmatrix} 1 \\ 1 \\ 1 \end{pmatrix}; \vec{e_2} = \begin{pmatrix} -1 \\ 1 \\ 0 \end{pmatrix}; \vec{e_3} = \begin{pmatrix} -1 \\ 0 \\ 1 \end{pmatrix}$$

And in this case, there is no problem at all, because we got equal number of eigenvalues and linearly independent eigenvectors. There are many reasons why this is important, however, probably one of the most common ones is this.

MATRIX DIAGONALIZATION

Matrix diagonalization is an advantageous and simple process. It is a special decomposition of matrices. However, it has applications in many areas and allows the user to do, understand, work and use eigenvalues and eigenvectors to reach higher goals. We will use a simple process to diagonalize a matrix. Suppose we have a matrix ***A*** with known eigenvalues and eigenvectors. We know that

$$A\vec{e_1} = \lambda_1 \vec{e_1}$$

But let us take it to the next level. Let us suppose we multiply ***A*** times a matrix formed by the eigenvectors and let us call it matrix ***S***. Therefore, this happens:

$$S = \left(\vec{e_1}, \vec{e_2}, \ldots, \vec{e_n}\right)$$

$$A \cdot S = A \cdot \left(\vec{e_1}, \vec{e_2}, \ldots, \vec{e_n}\right) = \left(\lambda_1 \vec{e_1}, \lambda_2 \vec{e_2}, \ldots, \lambda_n \vec{e_n}\right)$$

You can see that the right hand side is a matrix formed by all the eigenvectors multiplied by their respective eigenvalue. As we have each column multiplied by a constant, we can factor them in a diagonal matrix to the right. Therefore,

$$A \cdot S = \left(\vec{e_1}, \vec{e_2}, \ldots, \vec{e_n}\right) \cdot \begin{pmatrix} \lambda_1 & 0 & \cdots & 0 \\ 0 & \lambda_2 & \cdots & 0 \\ \vdots & \vdots & \cdots & \vdots \\ 0 & 0 & \cdots & \lambda_n \end{pmatrix}$$

The diagonal matrix that appears on the right is formed by all the eigenvalues of ***A***, and it is usually denoted by the Greek letter ***Λ***. Therefore, we can express the above system like this:

$$A \cdot S = S \cdot \Lambda$$

Wherein, as we have said before, ***S*** is a matrix formed by the eigenvectors of ***A***. Therefore, if we have ***n*** different linearly independent eigenvectors, then ***S*** is a square invertible matrix and we can do the following:

$$A = S \Lambda S^{-1}$$

which is another possible decomposition of A called eigendecomposition. It is very useful in many applications because it presents special properties. For example, remember in the preceding chapters when we talked about diagonalization and the powers of ***A***? This is it, because now we have an easy way to compute the powers of ***A***. For example, let us see what happens if we calculate the square of ***A*** by the above decomposition:

$$A^2 = \left(S \Lambda S^{-1}\right)\left(S \Lambda S^{-1}\right) = S \Lambda \left(S^{-1} \cdot S\right) \Lambda S^{-1} = S \Lambda^2 S^{-1}$$

If we apply the same methodology to the ***n^th^*** power of ***A***, we get:

$$A^n = S \Lambda^n S^{-1}$$

Which takes us somehow to the starting point, because we now need to calculate the ***n^th^*** power of ***Λ***. But the advantage here is that ***Λ*** is a diagonal matrix, therefore, calculating its power is just "powering" every component. To demonstrate how to do this we could check an example with the eigenvalues and eigenvectors we saw before. Let us use again this matrix:

$$B = \begin{pmatrix} 1 & 0 & 4 \\ 0 & 2 & 0 \\ 3 & 1 & -3 \end{pmatrix}$$

whose eigenvalues are 2, 3 and -5 and its eigenvectors are

$$\vec{e_1} = \begin{pmatrix} 4 \\ -7 \\ 1 \end{pmatrix}; \vec{e_2} = \begin{pmatrix} 2 \\ 0 \\ 1 \end{pmatrix}; e_3 = \begin{pmatrix} -2/3 \\ 0 \\ 1 \end{pmatrix}$$

Therefore, as our matrix has all different eigenvalues, then we can say that:

$B \cdot S = S\Lambda$

$$\begin{pmatrix} 1 & 0 & 4 \\ 0 & 2 & 0 \\ 3 & 1 & -3 \end{pmatrix} \cdot \begin{pmatrix} 4 & 2 & -2/3 \\ -7 & 0 & 0 \\ 1 & 1 & 1 \end{pmatrix} = \begin{pmatrix} 4 & 2 & -2/3 \\ -7 & 0 & 0 \\ 1 & 1 & 1 \end{pmatrix} \cdot \begin{pmatrix} 2 & 0 & 0 \\ 0 & 3 & 0 \\ 0 & 0 & -5 \end{pmatrix}$$

wherein we can see that the eigenvectors and eigenvalues are not written in a stochastic way -- the first eigenvalue corresponds to the first eigenvector and so on and so forth. From this, if we invert matrix ***S*** on the left side, we will see that matrix ***B*** is:

$$\begin{pmatrix} 1 & 0 & 4 \\ 0 & 2 & 0 \\ 3 & 1 & -3 \end{pmatrix} = \begin{pmatrix} 4 & 2 & -2/3 \\ -7 & 0 & 0 \\ 1 & 1 & 1 \end{pmatrix} \cdot \begin{pmatrix} 2 & 0 & 0 \\ 0 & 3 & 0 \\ 0 & 0 & -5 \end{pmatrix} \cdot \begin{pmatrix} 0 & -1/7 & 0 \\ 3/8 & 1/4 & 1/4 \\ -3/8 & -3/28 & 3/4 \end{pmatrix} = B$$

The value of ***B*** squared for example, would be:

$$B^2 = \begin{pmatrix} 4 & 2 & -2/3 \\ -7 & 0 & 0 \\ 1 & 1 & 1 \end{pmatrix} \cdot \begin{pmatrix} 2^2 & 0 & 0 \\ 0 & 3^2 & 0 \\ 0 & 0 & (-5)^2 \end{pmatrix} \cdot \begin{pmatrix} 0 & -1/7 & 0 \\ 3/8 & 1/4 & 1/4 \\ -3/8 & -3/28 & 3/4 \end{pmatrix} = \begin{pmatrix} 16 & 4 & -8 \\ 0 & 4 & 0 \\ -6 & -1 & 21 \end{pmatrix}$$

This is an example with different eigenvalues, which is a nice case, but let us check out other examples where we have repeated eigenvalues, like the matrix ***E*** that we saw before:

$$E = \begin{pmatrix} 1 & -1 & -1 \\ -1 & 1 & -1 \\ -1 & -1 & 1 \end{pmatrix}$$

Notice that its eigenvalues are 2 and -1, and that its eigenvectors are:

$$\vec{e_1} = \begin{pmatrix} 1 \\ 1 \\ 1 \end{pmatrix}; \vec{e_2} = \begin{pmatrix} -1 \\ 1 \\ 0 \end{pmatrix}; \vec{e_3} = \begin{pmatrix} -1 \\ 0 \\ 1 \end{pmatrix}$$

wherein the last two eigenvectors correspond to lambda equal to 2. Therefore, having this information we can express our matrix ***E*** like:

$$E \cdot S = S \cdot \Lambda$$

$$E = \begin{pmatrix} 1 & -1 & -1 \\ 1 & 1 & 0 \\ 1 & 0 & 1 \end{pmatrix} \begin{pmatrix} -1 & 0 & 0 \\ 0 & 2 & 0 \\ 0 & 0 & 2 \end{pmatrix} \cdot \frac{1}{3} \cdot \begin{pmatrix} 0 & 0 & 1 \\ -1 & 2 & -1 \\ -1 & -1 & 2 \end{pmatrix}$$

The third is a constant that belongs to the inverse of ***S.*** We can see what we have said before that even if we have repeated eigenvalues we will be able to diagonalize the matrix if we have ***n*** linearly independent eigenvectors, being ***n*** the size of the matrix we are diagonalizing. This is important, because there will be cases where we are not going to be able to diagonalize a matrix.

It is also important to notice that this also happens even to matrices that are not full rank (in that case, at least one eigenvalue is zero because the determinant of the matrix is already zero and has a null space). But before going further on this chapter, it is important to ***demonstrate two properties*** of any matrix that are useful in checking the eigenvalues.

TWO PROPERTIES OF EIGENVALUES

The first property is that "the product of all the eigenvalues is equal to the determinant of the matrix." This property comes from the following:

Suppose we have a matrix **A** of size **n**, therefore, it has **n** eigenvalues (does not matter whether they are real or complex, different or repeated). From this approach:

$$det(A-\lambda I)=P_n(\lambda)=(-1)^n\left[\lambda^n+c_1\lambda^{n-1}+\ldots+c_n\right]$$

From the fundamental theorem of algebra, we know that we can factor the polynomial depending on its roots just like this:

$$det(A-\lambda I)=P_n(\lambda)=(-1)^n(\lambda-\lambda_1)\cdot(\lambda-\lambda_2)\cdot\ldots\cdot(\lambda-\lambda_n)$$

which is a polynomial that will depend on variable lambda. If we set lambda equal to zero then:

$$det(A-0\cdot I)=P_n(0)=(-1)^n(0-\lambda_1)\cdot(0-\lambda_2)\cdot\ldots\cdot(0-\lambda_n)$$

$$det(A)=(-1)^n\cdot(-1)^n\cdot\lambda_1\cdot\lambda_2\cdot\ldots\cdot\lambda_n=\prod_{i=1}^{n}\lambda_i$$

From this result we can see that if we get to compute the eigenvalues correctly, then we already have the determinant of the matrix. This also supports the fact that if **A** is singular then one of the eigenvalues is zero by the way.

The other property of eigenvalues is that ***their sum is equal to trace of matrix A***. The trace is a very simple concept in matrices -- it is the sum of the coefficients of the main diagonal of the matrix. This can be proven by the following way, but first let us look at how to calculate the trace, and a property of the trace.

The trace of **A** can be calculated like this:

$$tr(A)=\sum_{i=1}^{n}a_{ii}$$

Now let us suppose **A** is the multiplication of two matrices **B** of size $n\times m$ and **C** of size $m\times n$, therefore, we can write the above like the following way:

$$tr(A)=\sum_{i=1}^{n}\sum_{j=1}^{m}b_{ij}\cdot c_{ji}$$

Expanding the sums and rearranging,

$$tr(A)=\begin{array}{c}(b_{11}\cdot c_{11}+b_{12}\cdot c_{21}+\ldots+b_{1m}\cdot c_{m1})\\ +(b_{21}\cdot c_{12}+b_{22}\cdot c_{22}+\ldots+b_{2m}\cdot c_{m2})\\ +\vdots\\ +(b_{n1}\cdot c_{1n}+b_{n2}\cdot c_{2n}+\ldots+b_{nm}\cdot c_{mn})\end{array}$$

$$tr(A)=\begin{array}{c}(c_{11}\cdot b_{11}+c_{12}\cdot b_{21}+\ldots+c_{1n}\cdot b_{n1})\\ +(c_{21}\cdot b_{12}+c_{22}\cdot b_{22}+\ldots+c_{2n}\cdot b_{n2})\\ +\vdots\\ +(c_{m1}\cdot b_{1m}+c_{m2}\cdot b_{2m}+\ldots+c_{mn}\cdot b_{nm})\end{array}$$

$$tr(A)=\sum_{i=1}^{n}\sum_{j=1}^{m}c_{ji}\cdot b_{ij}$$

We can see that we can invert the order of matrices while calculating the trace, therefore,

$$tr(A)=tr(B\cdot C)=tr(C\cdot B)$$

Now, suppose also that ***A*** can be diagonalized in terms of its eigenvalues and eigenvectors, therefore,

$$tr(A)=tr(S\ddot{E}\cdot S^{-1})=tr(S\,S^{-1}\ddot{E})=tr(\ddot{E})$$

In which we used the property we just discussed about changing the order of matrices, and as matrix Λ only has the eigenvalues of ***A***, then the sum of the eigenvalues is equal to the trace of ***A***. This property must be fulfilled always and is a way to check the eigenvalues of ***A***.

REPEATED EIGENVALUES

Now that we have reached this point, we can confront the situation in which there are repeated eigenvalues in a matrix and there are not enough eigenvectors to diagonalize it. Usually, many books and authors have the tradition to explain this through two concepts, which are the algebraic and geometric multiplicities.

These two names may sound special but in terms of eigenvalues and eigenvectors, its the following: ***the algebraic multiplicity*** is the number of times an eigenvalue is repeated in a set of eigenvalues, and ***the geometric***

multiplicity is the number of vectors that come out from an eigenvalue. For example, looking back at one of the matrices we saw before:

$$E = \begin{pmatrix} 1 & -1 & -1 \\ -1 & 1 & -1 \\ -1 & -1 & 1 \end{pmatrix}$$

Its characteristic polynomial is

$$P(\lambda) = (2-\lambda)^2(1+\lambda)$$

Therefore, eigenvalue 2 has algebraic multiplicity equal to 2 and eigenvalue -1 has algebraic multiplicity equal to 1. Now both eigenvalues give a certain number of eigenvectors. For example, eigenvalue 2 gives two vectors which are

$$\vec{e_2} = \begin{pmatrix} -1 \\ 1 \\ 0 \end{pmatrix}; \vec{e_3} = \begin{pmatrix} -1 \\ 0 \\ 1 \end{pmatrix}$$

Therefore, it has geometric multiplicity equal to 2. However, eigenvalue -1 gives only the following vector:

$$\vec{e_1} = \begin{pmatrix} 1 \\ 1 \\ 1 \end{pmatrix}$$

Therefore, has geometric multiplicity equal to 1. From here we can see that if the ***algebraic multiplicity is equal to geometric multiplicity*** then we have ***n*** different linearly independent eigenvectors and our matrix can be diagonalized. However, if that is not true, then we cannot diagonalize the matrix in terms of its eigenvalues and eigenvectors.

Jordan Form

However, this did not stop ***Camille Jordan*** during the second half of the nineteenth century to develop an algorithm to get very close to diagonalizing every matrix, whether it has full linearly independent eigenvectors or not. The algorithm is called ***Jordan canonical form*** and it is a generalization of what we have just seen.

Jordan canonical form is very similar with what we just saw but many of you will be able to notice the main differences. Jordan canonical form first calculates the eigenvalues and then the eigenvectors of a matrix. However, if the multiplicities are not the same for all eigenvalues, then he tricks the system like the following way. He decomposes matrix **A** into the following form:

$$AP = PJ$$

Where **P** is a matrix calculated accordingly to the eigenvectors, but matrix **J** is the main difference from what we have studied now since it is of the following form:

$$J = \begin{pmatrix} \lambda_1 & 1 & 0 & \cdots & 0 \\ 0 & \lambda_1 & 0 & \cdots & 0 \\ 0 & 0 & \lambda_2 & \cdots & 0 \\ \vdots & \vdots & \vdots & \cdots & \vdots \\ 0 & 0 & 0 & \cdots & \lambda_n \end{pmatrix}$$

where the eigenvalues are on the main diagonal, like usual eigendecomposition. However, above the repeated eigenvalues goes a 1 that does not make the matrix diagonal but it does make it as close as possible to the diagonal. There is a reason for getting ones over the repeated eigenvalues. Suppose we have a finite matrix **P** that has column vectors from $\boldsymbol{p_1}$ to $\boldsymbol{p_4}$, and that we have the following matrix **J** with 4 eigenvalues from which 2 are repeated:

$$J = \begin{pmatrix} \lambda_1 & 0 & 0 & 0 \\ 0 & \lambda_2 & 0 & 0 \\ 0 & 0 & \lambda_3 & 1 \\ 0 & 0 & 0 & \lambda_3 \end{pmatrix}$$

From our main equation, we can rewrite it as

$$A\left(\overrightarrow{p_1}, \overrightarrow{p_2}, \ldots, \overrightarrow{p_n}\right) - \left(\overrightarrow{p_1}, \overrightarrow{p_2}, \ldots, \overrightarrow{p_n}\right)J = 0$$

Which gives the following equations:

$$(A-\lambda_1 I)\overrightarrow{p_1}=\vec{0}$$
$$(A-\lambda_2 I)\overrightarrow{p_2}=\vec{0}$$
$$(A-\lambda_3 I)\overrightarrow{p_3}=\vec{0}$$
$$(A-\lambda_3 I)\overrightarrow{p_4}=\overrightarrow{p_3}$$

We can see that the first 3 equations correspond to the eigenvectors, but the last one does not, nevertheless, an arrangement is done. Suppose we multiply both sides of the last equation by $(A-\lambda_3 I)$:

$$(A-\lambda_1 I)\overrightarrow{p_1}=\vec{0}$$
$$(A-\lambda_2 I)\overrightarrow{p_2}=\vec{0}$$
$$(A-\lambda_3 I)\overrightarrow{p_3}=\vec{0}$$
$$(A-\lambda_3 I)^2\overrightarrow{p_4}=(A-\lambda_3 I)\overrightarrow{p_3}=0$$

Therefore, $\boldsymbol{p_4}$ is not an eigenvector strictly speaking, it is called a ***generalized eigenvector***. Remember that when finding the column vectors of matrix ***P*** we need to assure that all of them are linearly independent since we are going to have invert matrix ***P*** to decompose ***A***.

Jordan normal form has an extensive theory and terminology like the Jordan blocks, extending the kernel and many other things, that is, Jordan normal form has an immense interpretation under the light of abstract and linear algebra. However, it is an arrangement of data very close to an eigendecomposition. Let us do an example of a Jordan canonical form.

Suppose we have the following matrix ***A*** and we want to diagonalize it

$$A=\begin{pmatrix}5 & 4 & 2 & 1\\ 0 & 1 & -1 & -1\\ -1 & -1 & 3 & 0\\ 1 & 1 & -1 & 2\end{pmatrix}$$

If we calculate the characteristic polynomial we will see that it is $P(\lambda)=(\lambda-1)\cdot(\lambda-2)\cdot(\lambda-4)^2$

From this we can calculate the eigenvectors, hoping that we can get four linearly independent eigenvectors. And so the eigenvectors are:

$$for\lambda = 1 \rightarrow \vec{e_1} = \begin{pmatrix} -1 \\ 1 \\ 0 \\ 0 \end{pmatrix} \quad for\lambda = 2 \rightarrow \vec{e_2} = \begin{pmatrix} -1 \\ 1 \\ 0 \\ -1 \end{pmatrix} \quad for\lambda = 4 \rightarrow \vec{e_4} = \begin{pmatrix} 1 \\ 0 \\ -1 \\ 1 \end{pmatrix}$$

We can see here that we have three linearly independent vectors and not four. Therefore, our matrix ***is not diagonalizable, but it has a Jordan canonical form.*** Starting from the theory we just saw (which exactly corresponds to a case like this), we see that the eigenvalue that is causing trouble is 4, then we do the following. This is matrix $A - \lambda_4 I$:

$$A - \lambda_4 I = \begin{pmatrix} 1 & 4 & 2 & 1 \\ 0 & -3 & -1 & -1 \\ -1 & -1 & -1 & 0 \\ 1 & 1 & -1 & -2 \end{pmatrix}$$

And this is its square:

$$(A - \lambda_4 I)^2 = \begin{pmatrix} 0 & -9 & -5 & -5 \\ 0 & 9 & 5 & 5 \\ 0 & 0 & 0 & 0 \\ 0 & 0 & 4 & 4 \end{pmatrix}$$

To decompose $\boldsymbol{A}$ into a Jordan form we need to calculate the null space of the above matrix. By doing this we will see that it has two vectors in the null space. Let us call them $\boldsymbol{f}$.

$$\vec{f_1} = \begin{pmatrix} 1 \\ 0 \\ 0 \\ 0 \end{pmatrix}; \vec{f_2} = \begin{pmatrix} 0 \\ 0 \\ -1 \\ 1 \end{pmatrix}$$

These two vectors form a basis for one of the vectors that will be in our matrix $\boldsymbol{P}$. Specifically, they will be a basis for vector $\boldsymbol{p_4}$, so from here we need to find a linear combination of them such that:

$$(A - \lambda_4 I)\vec{p_4} = \vec{p_3}$$

Therefore, if we have two scalars ***a*** and ***b*** we get:

$$(A-\lambda_4 I)\cdot\left(a\cdot\vec{f_1}+b\cdot\vec{f_2}\right)=\vec{p_3}$$

And from here we will see that ***a***= 1 and ***b*** = 0. This linear combination represents our vector $\boldsymbol{p_4}$ that goes on our matrix ***P***. We could have solved the problem by solving for each component of $\boldsymbol{p_4}$ like this:

$$(A-\lambda_4 I)\vec{p_4}=\vec{p_3}$$

$$\begin{pmatrix} 1 & 4 & 2 & 1 \\ 0 & -3 & -1 & -1 \\ -1 & -1 & -1 & 0 \\ 1 & 1 & -1 & -2 \end{pmatrix}\begin{pmatrix} p_{41} \\ p_{42} \\ p_{43} \\ p_{44} \end{pmatrix}=\begin{pmatrix} 1 \\ 0 \\ -1 \\ 1 \end{pmatrix}$$

and choosing a nice value of vector $\boldsymbol{p_4}$. However, using this approach we have 4 variables (variables equal to the size of the column vectors). To find those variables we have to do row reduction with the matrix and the independent term which, numerically speaking, can cause trouble. These problems can arise because of numerical errors in computers, and we could get a row of zeros equal to a number when in reality that would not happen.

It is better to find the powers of the above matrix and get its null space (like we did) because this is far more stable, numerically speaking, and reduces the number of variables by a large amount. As we saw in our example, we found two vectors in the null space of $(A-\lambda_4 I)^2$ and we know that a linear combination of them is the vector that we are after. This approach reduces the number of variables because our variables are just the scalars that linearly combine the vectors in the null space of $(A-\lambda_4 I)^2$. Besides, calculating a null space is one of the easiest thing for computers and there is no danger of getting a "no solution" situation when, in reality, there is none.

Also, matrices of the form $(A-\lambda I)$ tend to be nilpotent. This means that as you get the powers of these matrices the result will at some point be the zero matrix. However, this is not always true. There are cases when they are not nilpotent but the possibility of them being nilpotent is highly valuable, because it is far easier to work with them and get their null spaces.

Now that we have discussed this, we have finished the problem, since we can build matrix ***P*** and ***J***. Remember that ***J*** is the matrix containing the eigenvalues, and ***P*** is the matrix containing the eigenvectors and generalized eigenvectors.

$$J = \begin{pmatrix} 1 & 0 & 0 & 0 \\ 0 & 2 & 0 & 0 \\ 0 & 0 & 4 & 1 \\ 0 & 0 & 0 & 4 \end{pmatrix}; P = \begin{pmatrix} -1 & -1 & 1 & 1 \\ 1 & 1 & 0 & 0 \\ 0 & 0 & -1 & 0 \\ 0 & -1 & 1 & 0 \end{pmatrix}$$

From here we can also find the inverse of ***P***:

$$P^{-1} = \begin{pmatrix} 0 & 1 & 1 & 1 \\ 0 & 0 & -1 & -1 \\ 0 & 0 & -1 & 0 \\ 1 & 1 & 1 & 0 \end{pmatrix}$$

And we now have that matrix ***A*** can be rewritten as:

$$A = PJP^{-1} = \begin{pmatrix} -1 & -1 & 1 & 1 \\ 1 & 1 & 0 & 0 \\ 0 & 0 & -1 & 0 \\ 0 & -1 & 1 & 0 \end{pmatrix} \cdot \begin{pmatrix} 1 & 0 & 0 & 0 \\ 0 & 2 & 0 & 0 \\ 0 & 0 & 4 & 1 \\ 0 & 0 & 0 & 4 \end{pmatrix} \cdot \begin{pmatrix} 0 & 1 & 1 & 1 \\ 0 & 0 & -1 & -1 \\ 0 & 0 & -1 & 0 \\ 1 & 1 & 1 & 0 \end{pmatrix}$$

Which is not a diagonal decomposition, but very close indeed.

SIMILAR MATRICES

An interesting fact about this kind of decompositions is that they are the definition of similar matrices. Similar matrices are a simple concept that can be sometimes useful depending on the user's insight. Let us take a look at the theory.

Two matrices, ***A*** and ***B***, are called similar if there exists a matrix ***M*** such that:

$$A = MBM^{-1}$$

And the great fact about two matrices being similar is that they have the same eigenvalues (not necessarily the same eigenvectors, but yes the eigenvalues) and this is because

$A\vec{x} = \lambda\vec{x}$

We can input A in terms of matrix B

$MBM^{-1}\vec{x} = \lambda\vec{x}$

And multiplying at both sides by M^{-1} and rearranging we get

$B(M^{-1}\vec{x}) = \lambda(M^{-1}\vec{x})$

Which can be rewritten as

$B\vec{v} = \lambda\vec{v}$

Therefore, matrices ***A*** and ***B*** have the same eigenvalues, but not necessarily the same eigenvectors.

THE POWERS OF A

Suppose we have a matrix ***A*** with all different eigenvalues and linearly independent eigenvectors. We are interested to see what happens when we compute ***A*** to the ***nth*** power. Three main things can happen. Let us take a look at those.

We know that the ***nth*** power of ***A*** can be calculated like:

$A^n = S\Lambda^n S^{-1}$

Matrix ***S*** and its inverse are important but they don't change as ***n*** grows. What we have to look at is the eigenvalues.

$$\Lambda^n = \begin{pmatrix} \lambda_1^n & 0 & \cdots & 0 \\ 0 & \lambda_2^n & \cdots & 0 \\ \vdots & \vdots & \cdots & \vdots \\ 0 & 0 & \cdots & \lambda_m^n \end{pmatrix}$$

We see that as ***n*** grows to infinity, three things can happen:

- if all the eigenvalues are between -1 and 1, then their powers will converge to zero as ***n*** grows to infinity;. Therefore, matrix ***A*** is nilpotent.
- if all the eigenvalues are one, then ***A*** does not change as ***n*** grows to infinity. In this case, $A^n = A$.

- if at least one of the eigenvalues has absolute value greater than one, then we can assure that as ***n*** grows to infinity then A^n grows to infinity as well.

However, it does not fulfill for matrices with Jordan form. We just mentioned these cases because it is good to be able to check the eigenvalues and check for convergence in case we are solving systems of equations numerically. This is because iterative methods must converge, and this is a way of assuring convergence. This will be better understood later on. For now, let us check certain example of matrices that fulfill this property. One of these is the ***Markov matrix***.

MARKOV MATRIX

A Markov matrix is a matrix whose rows or columns add to one and every element in the matrix is greater or equal to zero. It is assured that a Markov matrix has one eigenvalue equal to one (easy to prove) and all the other eigenvalues have absolute value less than one. For example, a Markov matrix would look like this:

$$A = \begin{pmatrix} 0.2 & 0.4 & 0.7 & 0.1 \\ 0.4 & 0.4 & 0.3 & 0.21 \\ 0.3 & 0.1 & 0 & 0.39 \\ 0.1 & 0.1 & 0 & 0.3 \end{pmatrix}$$

Since all elements are equal to or greater than zero and all columns add to 1, it is assured that from these conditions all the eigenvalues have absolute value less than 1. Therefore, it is almost a nilpotent matrix (it is not nilpotent because of that eigenvalue equal to 1). Markov matrices have extensive use along sciences, and serve well to model probability problems, financial situations and obviously linear algebra. For example, one simple application that will help us understand not only Markov matrices but also an introduction to numerical methods is the population movement.

To make it easy, let us do a 2 by 2 system. Suppose we have two cities ***A*** and ***B***, and people are constantly moving from one city to another. Statisticians say that there is an 80% probability on a yearly basis that a person in city ***A*** will move to city ***B***, and that there is a 60% probability on a yearly basis that a person in city ***B*** will move to city ***A***, but they want to simulate how many people are in each city at a given time.

We can express this problem in the following way. Let us suppose we have a vector ***p*** that represents the people in city ***A*** and in city ***B***, then:

$$\vec{p} = \begin{pmatrix} p_A \\ p_B \end{pmatrix}$$

So applying our conditions we can simulate it like this:

$$\overrightarrow{p_{k+1}} = C\overrightarrow{p_k}$$

Where matrix ***C*** is a probability matrix and vector $\boldsymbol{p_k}$ are the people in each city at a given time. By the way, we have all the information we need to solve the problem, because from what we are told we can also conclude that 20% of the people in city ***A*** will stay in city ***A*** and that 40% of the people in city ***B*** will stay in city ***B***. Suppose that initial conditions are the following:

$$\overrightarrow{p_0} = \begin{pmatrix} 5000 \\ 30000 \end{pmatrix}$$

Therefore, our system is:

$$\overrightarrow{p_k} = \begin{pmatrix} 0.2 & 0.6 \\ 0.8 & 0.4 \end{pmatrix}^k \overrightarrow{p_0}$$

After a year, we get that the number of people in each city is:

$$\overrightarrow{p_1} = \begin{pmatrix} 0.2 & 0.6 \\ 0.8 & 0.4 \end{pmatrix}\begin{pmatrix} 5000 \\ 30000 \end{pmatrix} = \begin{pmatrix} 19000 \\ 16000 \end{pmatrix}$$

And after two years, we get that the number of people in each city is:

$$\overrightarrow{p_2} = \begin{pmatrix} 0.2 & 0.6 \\ 0.8 & 0.4 \end{pmatrix}\begin{pmatrix} 19000 \\ 16000 \end{pmatrix} = \begin{pmatrix} 13400 \\ 21600 \end{pmatrix}$$

But after many years, assuming that there is no change in the system… how many people would be in each city? To do this (easily) we need to find the eigenvalues and eigenvectors of ***C***. This is easy as we don't even have to calculate the determinant. We already know that one eigenvalue is 1 (because it is a Markov matrix) and that the sum of the eigenvalues (in this case we have two eigenvalues) must be equal to the trace (in this case 0.6). Therefore, our two eigenvalues are 1 and -0.4. Calculating the eigenvectors gives us

$$\vec{e_1} = \begin{pmatrix} 3 \\ 4 \end{pmatrix}; \vec{e_2} = \begin{pmatrix} 1 \\ -1 \end{pmatrix}$$

And from here we can transform our matrix ***C*** into:

$$\vec{p_k} = \frac{1}{7}\begin{pmatrix} 3 & 1 \\ 4 & -1 \end{pmatrix} \cdot \begin{pmatrix} 1 & 0 \\ 0 & -0.4 \end{pmatrix}^{k \cdot} \begin{pmatrix} 1 & 1 \\ 4 & -3 \end{pmatrix} \cdot \begin{pmatrix} 5000 \\ 30000 \end{pmatrix}$$

and if we make ***k*** approach infinity we will get

$$\vec{p_\infty} = \frac{1}{7}\begin{pmatrix} 3 & 1 \\ 4 & -1 \end{pmatrix} \cdot \begin{pmatrix} 1 & 0 \\ 0 & 0 \end{pmatrix} \cdot \begin{pmatrix} 1 & 1 \\ 4 & -3 \end{pmatrix} \cdot \begin{pmatrix} 5000 \\ 30000 \end{pmatrix}$$

Thus, calculating the value of vector ***p*** will give us

$$\vec{p_\infty} = \begin{pmatrix} 15000 \\ 20000 \end{pmatrix}$$

Notice that the total amount of people in both cities is always 35000, and that sum has been preserved in all the calculations we made.

SYMMETRIC MATRICES

Symmetric matrices are very easy to work with and, as we may have noticed in several examples and properties, are very desirable. Under the light of eigenvectors and eigenvalues they are still desirable for two reasons. ***The first reason*** is that all the eigenvalues of a symmetric matrix are real, non-imaginary numbers.

To prove this, let us take the following expression. Assume ***A*** is a symmetric matrix of real coefficients, and suppose it has an eigenvalue lambda which can be complex, and an eigenvector ***x*** that can be also complex, therefore,

$$A\vec{x} = \lambda\vec{x}$$

If we conjugate both sides of the equation we would get

$$\overline{A}\overline{\vec{x}} = \overline{\lambda}\overline{\vec{x}}$$

But as matrix $\boldsymbol{A}$ is a real coefficient matrix then its conjugate is equal to $\boldsymbol{A}$. Now if we transpose the last expression we got we would get the following. Remember that lambda is a number therefore, its transpose is itself, and we assumed that matrix $\boldsymbol{A}$ is symmetrical therefore, its transpose is still $\boldsymbol{A}$:

$$\overline{\vec{x^T}}A = \overline{\lambda}\overline{\vec{x^T}}$$

And if we multiply the above expression by vector $\boldsymbol{x}$ on the right we will get

$$\overline{\vec{x}^T}A\vec{x} = \overline{\lambda}\overline{\vec{x}^T}\vec{x}$$

But we now that as $\boldsymbol{x}$ is an eigenvector of $\boldsymbol{A}$, then matrix $\boldsymbol{A}$ times vector $\boldsymbol{x}$ is equal to lambda times vector $\boldsymbol{x}$. Therefore,

$$\overline{\vec{x}^T}\lambda\vec{x} = \overline{\lambda}\overline{\vec{x}^T}\vec{x}$$

and as lambda is just a number, then we can put it this way

$$\lambda\overline{\vec{x}^T}\vec{x} = \overline{\lambda}\overline{\vec{x}^T}\vec{x}$$

and from here we can conclude that

$$\overline{\lambda} = \lambda$$

and the only way for the conjugate of lambda to be equal to lambda is if lambda is real.

Vector's Norm

From this proof we can also see one thing which we still have not worked with, but is really important in other fields of algebra. We just saw this expression:

$$\overline{\vec{x}^T}\vec{x}$$

From past knowledge we know that $x^T \cdot x$ *is equal to the length squared of* $\boldsymbol{x}$*. But in case* $\boldsymbol{x}$ *is a complex number, then its length squared is calculated like this* $\overline{\vec{x}^T}\vec{x}$*, because the result will be a real number. Let us check this out. Suppose vector* $\boldsymbol{x}$ *has the following components:*

$$\vec{x} = \begin{pmatrix} a_1 + i \cdot b_1 \\ a_2 + i \cdot b_2 \\ \vdots \\ a_n + i \cdot b_n \end{pmatrix} = \begin{pmatrix} a_1 \\ a_2 \\ \vdots \\ a_n \end{pmatrix} + i \cdot \begin{pmatrix} b_1 \\ b_2 \\ \vdots \\ b_n \end{pmatrix} = \vec{a} + i\vec{b}$$

The complex conjugate transpose of ***x*** *will have the following form:*

$$\overline{\overrightarrow{x^T}} = \begin{pmatrix} a_1 - i \cdot b_1 & a_2 - i \cdot b_2 & \cdots & a_n - i \cdot b_n \end{pmatrix} = \begin{pmatrix} a_1 & a_2 & \cdots & a_n \end{pmatrix} - i \cdot \begin{pmatrix} b_1 & b_2 & \cdots & b_n \end{pmatrix} = \overrightarrow{a^T} - i\overrightarrow{b^T}$$

Therefore, if we multiply both expressions we will get

$$\overline{\overrightarrow{x^T}} \cdot \vec{x} = \left(\overrightarrow{a^T} - i\vec{b}\right) \cdot \left(\overrightarrow{a^T} + i\vec{b}\right) = \| a \|^2 + \| b \|^2 = \| x \|^2$$

This is not so important right now, but it is worth mentioning this because we need to embrace complex numbers, and that includes complex vectors and matrices also.

Going back to our topic, we proved that symmetric matrices have real eigenvalues, but we also have to prove that ***they have orthogonal eigenvectors***. Again, suppose we have a matrix ***A*** which is symmetrical, and suppose that we can eigendecompose it. Therefore,

$$A = S\Lambda S^{-1}$$

Let us transpose the above expression

$$A^T = \left(S\Lambda S^{-1}\right)^T = \left(S^{-1}\right)^T \cdot \Lambda^T \cdot S^T$$

But matrix ***A*** and ***Λ*** are both symmetrical matrices, so

$$A = \left(S^{-1}\right)^T \cdot \Lambda \cdot S^T$$

From this last expression we can see, if we compare it with our main expression, S^T must be equal to S^{-1}. The only way for this to happen is if ***S*** is an orthonormal matrix (which is denoted as ***Q*** in other chapters). Therefore, the eigendecomposition for a symmetrical matrix is of the form:

$$A = Q\Lambda Q^T$$

which is immensely useful and easy to work with, since to invert matrix ***Q*** we only need to transpose it. Aside from these two properties, symmetrical matrices may also have another great property if they are positive definite. Let us look at that property.

Suppose we have a positive definite symmetric matrix **A** with real entries, and as always, we also have the following expression:

$A\vec{x} = \lambda\vec{x}$

Let us multiply both sides by $\overrightarrow{x^T}$. Before writing the equation, we know that if **A** is a positive definite matrix then it admits a Cholesky's decomposition. Therefore,

$$\overrightarrow{x^T} A\vec{x} = \lambda\overrightarrow{x^T}\vec{x}$$

$$A = U^T \cdot U$$

$$\overrightarrow{x^T} U^T U\vec{x} = \lambda\overrightarrow{x^T}\vec{x}$$

$$(U\cdot\vec{x})^T \cdot (U\cdot\vec{x}) = \lambda\overrightarrow{x^T}\cdot\vec{x}$$

We know that lambda is real (therefore, the eigenvectors are also real), but lambda could be positive or negative. However, in this case, we will see something very particular. We know that matrix **U** times vector **x** is another vector, and that a vector times its transpose is the length squared of that vector. Therefore,

$$\| U\cdot\vec{x} \|^2 = \lambda\cdot\| \vec{x} \|^2$$

Now we see that both sides of the equation are positive quantities. Therefore, lambda must also be positive to respect the equality. It is important to notice that the above expression remains true for any eigenvector of matrix **A**, but another case is that lambda could also be equal to zero. This case corresponds if **A** is not a full rank matrix because a matrix that is not full rank has an eigenvalue equal to zero.

So, in conclusion, if **A** is positive definite with full rank, all of its eigenvalues are real positive numbers; and if it is not full rank, then at least one of its eigenvalues is zero. Showing an example of this is trivial since we already saw examples, but cases we can use this information to check procedures and results in the succeeding discussions. By the way, the same approach for these properties can be done for anti-symmetrical properties to show that all of their eigenvalues are purely imaginary.

EIGENVALUES AND EIGENVECTORS IN REAL LIFE

Now that we have seen a lot of examples, it would be nice talk about how to ***compute the eigenvectors and eigenvalues in real life***. First let us discuss a detail about... almost anything we have done so far in this book. In many situations and approaches we have made, theory states a problem and how to compute it, however, in practice we see that we rely on other theories to accomplish our goals.

For example, one of the most remarkable examples are determinants, since computing determinants by Laplace's method, even for a 4x4 matrix, is very difficult and time consuming. For this we rely on other properties, like the properties of determinants, and we discovered that an easier way is to reduce the matrix into a diagonal form and compute the determinant.

The main idea of what we just said is that theory is necessary to understand the world around us, however, raw and unrefined theory can be difficult to implement in our daily basis. For this we use other theories that complement the one we have to make it is easier to implement. You may have noticed that this is the case in the many examples that we had. Also, you may have noticed that we have got to this point because of a common effort of countless mathematicians, statisticians, physicists and more, that even if the goal seems hard to obtain they have kept going, approximating the solution by easier ways.

This is said because in real life, calculating eigenvalues and eigenvectors is not an easy task. Calculating a determinant of a 4x4 matrix is difficult enough, calculating the eigenvalues is harder. In reality, matrices are not only 4x4. They can be hundreds of rows and columns, probably thousands or even millions. Calculating this determinant by regular means, even for a computer would be nonsense, at least with computers nowadays. For this. we rely on other theories. For example, we saw that the trace of a matrix is equal to sum of the eigenvalues and that the determinant of a matrix is equal to the product of the eigenvalues. For the rest of this chapter we will look at other properties that will help us along the way. The first one of them is **Gershgorin circle theorem.**

Gershgorin Circle Theorem

Gershgorin circle theorem is a valuable theorem for computing eigenvalues. Let A be an $n \times n$ matrix with known coefficients. Let

$R_i = \sum_{i \neq j} |a_{ij}|$ be the absolute values of the non diagonal entries in the ith row.

Let $D(a_{ii}, R_i)$ be the closed disc centered at a_{ii} with radius R_i. Such a disc is called a Gershgorin disc. **Gershgorin circle theorem** states that every eigenvalue of A lies within at least one of the Gershgorin discs.

The proof of this theorem comes very easy. Suppose we have an eigenvalue lambda and an eigenvector ***v*** which we scaled to have the ith component equal to 1 and all the other components with absolute value less than 1. From our main equation we get

$$A\vec{x} = \lambda\vec{x} \rightarrow \sum_{j=1}^{n} a_{ij}x_j = \lambda x_i = \lambda$$

$$a_{ii}x_i + \sum_{j \neq i}^{n} a_{ij}x_j = \lambda$$

We know that x_i is 1, but let us rearrange and apply the triangle inequality

$$\left|\lambda - a_{ii}\right| = \left|\sum_{j \neq i} a_{ij}x_j\right| \leq \sum_{j \neq i} |a_{ij}||x_j|$$

and as all the x_j components are less than one

$$|\lambda - a_{ii}| \leq \sum_{j \neq i} |a_{ij}||x_j| \leq \sum_{j \neq i} |a_{ij}| = R_i$$

Therefore, this theorem gives us an idea of where the eigenvalues might be (whether they are real or complex). But how can this theorem be useful? When computing eigenvalues of large matrices we do not know where the eigenvalues are -- no clue -- but this theorem gives us a clue where they are. Usually when calculating eigenvalues we use iterative methods, but to start an iterative we need an initial condition or a seed. The better the choice of that initial condition is, the better results we can get and the faster we can get them.

An example of this theorem will be left for the next chapter because we will see how eigenvalues and eigenvectors can be calculated by other means.

CHAPTER 6

LINEAR ALGEBRA APPLICATIONS

"It is hard to convince a high school student that he will encounter a lot of problems more difficult than those of algebra and geometry"

E. W. Howe

INTRODUCTION

Until now we have seen, studied, analyzed and proved many theorems and procedures in linear algebra. However, the applications this has is almost limitless since every day new questions and problems arise that can be solved using a linear algebra approach. However, as time goes by many procedures and applications are left forgotten because new, better and easier algorithms are developed. Examples of this can be seen mostly in engineering since theories and algorithms are continually developing into higher and faster procedures, especially now with computers. Therefore, we will discuss in this chapter the most common and faster algorithms to solve problems in an easy and comfortable way.

Since we are going to use algorithms that require computational power, it is recommended that you look for programs or aid in computation so that you will be able to solve problems of this kind on your own. If you decide to make a program of your own algorithms, which, in some cases, would be the best, many programming languages like Fortran, C, C++ and Python are available. Having said this, all problems in this chapter will be similar to real-world problems. Therefore, systems and matrices on this chapter will be usually of size 4 or 5, with some exceptions.

MAIN APPLICATIONS

Matrices can enter distinct worlds around them, like the world of calculus in particular. For example, let us start with the most common case, numerical calculus. Numerical calculus is an area that tries to solve problems through arithmetic where differential calculus is not the best approach -- like when interpolating functions. Many methods are able to interpolate functions but the most common ones are Newton's approach and Lagrange's approach. Several studies have been done on these subjects and have well-supported basis. Some names should sound familiar to the reader, like Vandermonde's matrix, which is one of the most famous matrices in this area, as it is used to solve for coefficients when interpolating with polynomials.

Matrix Calculus

Even though this is relatively new in the world of mathematics, matrix calculus is a rising area nowadays. **Matrix calculus** is a science that relates differential calculus with matrices and vectors. This new subject studies the cases when deriving vectors with respect to a scalar, or when deriving vectors with respect to other vectors, or even when deriving matrices with respect to scalars or scalars with respect to matrices. Matrix calculus is one of the generalizations that come out in the mathematical world and allows us to study high end phenomena with a simple approach.

Physical Applications of Matrices

Matrices are found in many places through physics. From differential equations to Newton's law of motion. At first, when the fundamental equations that rule the motion and interaction of bodies were born they were proposed as scalar equations, which we could say are one dimensional. But as abstraction, knowledge and algebra became bigger, they have now evolved to vector equations and matrix notation has been added. The most amazing example is Newton's second law of motion which can be written as:

$$M\ddot{x} + B\dot{x} + Kx = F$$

where ***M***, ***B***, ***K*** and ***F*** are not scalars. They are matrices that represent the body of study. This is most seen in three dimensional problems, however, these are mostly solved using ***finite element analysis*** or ***finite difference approximations***. Other particular example of matrices along the physical world is the Inertia matrix, which is shown next:

$$I = \begin{pmatrix} I_{xx} & I_{xy} & I_{xz} \\ I_{yx} & I_{yy} & I_{yz} \\ I_{zx} & I_{zy} & I_{zz} \end{pmatrix}$$

Inertia matrix represents the inertia of a body of study. If the reader does not know, Inertia represents the resistance made by a body to rotate along with respect to a certain axis. This matrix usually is symmetrical, however, there are some cases in which this does not happen. That this matrix is symmetrical is very pleasing and convenient since it would reduce the amount of time on calculating its whole 9 components to only 6 components. This may not sound like a lot but it's important to say that all of those terms come from triple integration equations that will depend on the configuration of the body under study.

Differential Equations

Systems of linear differential equations are an excellent example of linear algebra coming together with other sciences, since systems of linear differential equations, whether they are homogeneous or not, are systems that can model life itself, and linear algebra helps to deal with these systems in an easier way. Matrix notation simplifies very much the difficulties to solve them. Not only that, systems of differential equations will have solutions based on the same principles we saw on past chapter, especially with an eigenvalue approach.

$$\dot{x}_1(t) = a_{11}(t) \cdot x_1(t) + a_{12}(t) \cdot x_2(t) + \ldots + a_{1n}(t) \cdot x_n(t)$$
$$\dot{x}_2(t) = a_{21}(t) \cdot x_1(t) + a_{22}(t) \cdot x_2(t) + \ldots + a_{2n}(t) \cdot x_n(t) \rightarrow \begin{pmatrix} \dot{x}_1(t) \\ \dot{x}_2(t) \\ \vdots \\ \dot{x}_n(t) \end{pmatrix}$$
$$\vdots$$
$$\dot{x}_n(t) = a_{n1}(t) \cdot x_1(t) + a_{n2}(t) \cdot x_2(t) + \ldots + a_{nn}(t) \cdot x_n(t)$$
$$= \begin{pmatrix} a_{11}(t) & a_{12}(t) & \cdots & a_{1n}(t) \\ a_{21}(t) & a_{22}(t) & \cdots & a_{2n}(t) \\ \vdots & \vdots & \cdots & \vdots \\ a_{11}(t) & a_{12}(t) & \cdots & a_{1n}(t) \end{pmatrix} \cdot \begin{pmatrix} x_1(t) \\ x_2(t) \\ \vdots \\ x_n(t) \end{pmatrix}$$

From the above, we can transform our $n \times n$ system of differential equations into the following:

$$\dot{\vec{x}}(t) = A(t) \cdot \vec{x}(t)$$

Vibrations and Oscillation

Vibrations and oscillating bodies are also examples of using matrix notation under calculus, especially when many bodies are connected together by some restriction. In these cases, when two bodies vibrate together they can only vibrate at very close frequencies which exactly correspond to the eigenvalues of the system of equations that correlates the two bodies. The usual equations are a set of ordinary differential equations like we just saw above.

Econometrics

Econometrics is the application of statistical methods into the world of economy. For this reason matrices populate the economic world as they are useful in representing many phenomena over time, like price variation, stock market, offers and more. If this is of interest to you, it is recommended that you look for a book on Econometrics and search for the different auto-regressive models, like AR, VAR, ARMA, ARIMA, and so on. That is a whole world on matrix application.

ITERATIVE METHODS FOR SOLVING SYSTEMS OF EQUATIONS

Now that we have seen all of these applications and examples, we need to see how to solve them one at a time. Let us start with the simplest field

-- solving systems of linear equations. At this point the reader is very familiar on how to solve systems of linear equations since that is one of the subjects we have already done. However, there exists other methods that are also useful in solving systems of equations. These methods, in some cases, are easier than applying Gaussian elimination, or factoring a matrix into an LU form, but some conditions must be met for these to be applied since these methods are iterative methods, and therefore, they do not always work.

In this section we will discuss four main methods: the ***Richardson's Method, Jacobi's Method, Gauss-Seidel Method and Successive Over Relaxation Method***. These methods are highly valuable today and correspond to another field of algebra which is numerical algebra. However, these methods will be explained here because many applications today may be solved by a numerical approach, besides it helps to have some knowledge of how these methods work. One of the applications that are most commonly solved with numerical methods are ***finite element analysis*** and we will see why.

Numerical Linear Algebra

Numerical linear algebra *is a branch of algebra that focuses on solving problems numerically rather than analytically, therefore, arrangements and other manipulations are done to equations to solve them by iterations given a certain algorithm. There are many methods in numerical linear algebra, methods that have been evolving since the invention of algebra. You are probably familiar with some methods to calculate roots of polynomials like bisection, regula falsi, Newton Raphson method, quasi Newton method, and Bairstow's method.*

In the last decades many methods have also appeared, but for calculating derivatives of complex functions, like Runge Kuta methods, or to calculate integrals of functions that are not integrable, like Simpson's Rule. But also in the last decades many energy has been invested in optimizing the existing methods and getting more stability during simulations.

Usually this approach has been done for two reasons, first to reduce computation time when calculating variables and iterative process, because this takes time, consumes electricity, memory and above all, an optimized method has less errors than other not-so-refined methods. The other reason is because of Moore's law. ***Moore's law*** *states that almost every two years the quantity of transistors in a microprocessor or a microchip are doubled. However, Moore's law is ending because transistors are getting smaller reaching almost molecular levels, therefore, microchips are reaching a limit, too.*

To solve that problem, optimization must be done. While this book is written, NASA and Google are working on quantum computers, which is a set of computers that have computation power far greater than regular computers.

Having said all of these, let us state the problem of numerical algebra. Suppose we have a system of equations of the following form:

$$A\vec{x} = \vec{b}$$

We have already seen this system countless times, however, we have only discussed the theory and small examples. In reality, theory and practice go together but sometimes it is impossible to apply theoretical methods to real examples, because real systems of equations can have more than a thousand of equations, or sometimes even a million of equations (like in electrical systems, where there are grids of thousands of homes, each one representing a variable). Imagine computing a the solution with such a matrix, and it is not only just that, it is also the fact that when subtracting quantities really close to each other the result may be so small that our computer may confuse it with a zero.

There is a joke in engineering, "paper is the strongest material on earth, paper can handle everything." In a piece of paper we can write a number of the following magnitude:

$$a = 1 \cdot 10^{-50}$$

Which is a very small number. Introducing that number into a computer would make the computer understand it as a zero, because even when problems and variables in theory are continuous, in reality phenomena and calculations are discrete to a certain level.

For example, until a few years ago, computers could only handle numbers with less than twenty decimal places, however, nowadays that number should have risen because of the latest advancements in computation.

To start talking about numerical methods we first need to understand two concepts besides the ones we already mentioned: ***convergence*** *and* ***rate of convergence****. Convergence, as we know, is a criteria that tells if a solution is approximating its real value or if it is not, and the rate of convergence is how fast the solution is approximating its value. To understand these two concepts, let us see two very simple examples.*

First, let us understand the concept about rearranging an equation. Suppose we have the following equation:

$$x = 2$$

We can arrange it in the following way:

$$x = \frac{x}{2} + 1$$

We can see that the solution to the equation is 2, however, let us assume that we don't know that. Numerical methods consist of getting a variable in terms of itself like what we just did: assume a first value of our variable, input it into our equation, get a new value to our variable and repeat the process. Let us show this with our last equation assuming our initial value of ***x*** *is 0. We denote this by saying* $\boldsymbol{x_0} = 0$*, therefore,*

$$x_1 = \frac{x_0}{2} + 1 = \frac{0}{2} + 1 = 1$$

We continue our process like I said

$$x_2 = \frac{x_1}{2} + 1 = \frac{1}{2} + 1 = 1,5$$

$$x_3 = \frac{x_2}{2} + 1 = \frac{1,5}{2} + 1 = 1,75$$

$$x_4 = \frac{x_3}{2} + 1 = \frac{1,75}{2} + 1 = 1,875$$

If we would continue iterating, then our solution would converge into a value, which in this case is 2. It converges because it is getting closer and closer to the solution, slowly but converging. Here comes the other important concept, which is the rate of convergence. There are methods that converge faster than others, depending on the right assumptions and optimization done to thc algorithm.

For now we will see how this numerical methods are proposed. We will look at some examples and explanations also. Going back to our original generic system of equations $A\vec{x}=\vec{b}$, the way to express the variable in terms of itself is usually done in the following way:

$$(A-Q+Q)\vec{x}=\vec{b}$$

$$Q\cdot\vec{x}=(Q-A)\vec{x}+\vec{b}$$

$$\overrightarrow{x_k}=Q^{-1}(Q-A)\overrightarrow{x_{k-1}}+Q^{-1}\cdot\vec{b}$$

$$\overrightarrow{x_k}=\left(I-Q^{-1}\cdot A\right)\overrightarrow{x_{k-1}}+Q^{-1}\cdot\vec{b}$$

Where Q can be any matrix (obviously invertible) and will depend on the choice of method. From our last expression we need to start iterating to see what happens. Supposing we start with an initial value of our vector (usually called the seed), then as we iterate our solutions would look like this:

$$\overrightarrow{x_1}=\left(I-Q^{-1}\cdot A\right)\overrightarrow{x_0}+Q^{-1}\cdot\vec{b}$$

$$\overrightarrow{x_2}=\left(I-Q^{-1}\cdot A\right)^2\overrightarrow{x_0}+\left(I-Q^{-1}\cdot A\right)\cdot Q^{-1}\cdot\vec{b}+Q^{-1}\vec{b}$$

$$\overrightarrow{x_3}=\left(I-Q^{-1}\cdot A\right)^3\overrightarrow{x_0}+\left(I-Q^{-1}\cdot A\right)^2\cdot Q^{-1}\cdot\vec{b}+\left(I-Q^{-1}\cdot A\right)\cdot Q^{-1}\vec{b}+Q^{-1}\cdot\vec{b}$$

$$\overrightarrow{x_n}=\left(I-Q^{-1}\cdot A\right)^n\overrightarrow{x_0}+\left(\sum_{i=0}^{n-1}\left(I-Q^{-1}\cdot A\right)^i\right)\cdot Q^{-1}\cdot\vec{b}$$

The above summation can be rewritten as:

$$\overrightarrow{x_n}=\left(I-Q^{-1}\cdot A\right)^n\overrightarrow{x_0}-A^{-1}\cdot Q\left[\left(I-Q^{-1}\cdot A\right)^n-I\right]\cdot Q^{-1}\cdot\vec{b}$$

This is the theory we need for an iterative method, but in practice we are not going to use an equation like the one above, because if you look closely, it involves the inverse of A. And if we would have the inverse of A, then why would we be doing this? Besides, the choice of matrix Q will make the convergence to our system faster or slower depending on the kind of system we are solving. For example, if we choose Q to be the identity matrix, then the above expression transforms into:

$$\overrightarrow{x_n} = (I - A)^n \overrightarrow{x_0} - A^{-1}\left[(I - A)^n - I\right] \cdot \vec{b}$$

*However, this is just an option. Any logical numerical method is acceptable. Some authors choose **Q** to be diagonal, others choose it triangular, other choose it with other properties -- choices are infinite. However, there is one condition: as **n** goes bigger (meaning more iterations) our solution should converge to a certain value, which means*

$$x_n - x_{n-1} \approx 0$$

Therefore, making the above subtraction

$$\overrightarrow{x_n} = (I - Q^{-1} \cdot A)^n \overrightarrow{x_0} - A^{-1} \cdot Q\left[(I - Q^{-1} \cdot A)^n - I\right] \cdot Q^{-1} \cdot \vec{b}$$

$$\overrightarrow{x_{n-1}} = (I - Q^{-1} \cdot A)^{n-1} \overrightarrow{x_0} - A^{-1} \cdot Q\left[(I - Q^{-1} \cdot A)^{n-1} - I\right] \cdot Q^{-1} \cdot \vec{b}$$

$$\overrightarrow{x_n} - \overrightarrow{x_{n-1}} = \left[(I - Q^{-1} \cdot A)^n - (I - Q^{-1} \cdot A)^{n-1}\right] \overrightarrow{x_0} - A^{-1} \cdot Q\left[(I - Q^{-1} \cdot A)^n - (I - Q^{-1} \cdot A)^{n-1}\right] \cdot Q^{-1} \cdot \vec{b} \approx \vec{0}$$

From here we can see that the above equation will have a solution if

$$(I - Q^{-1} \cdot A)^n \approx 0$$

Therefore, introducing the concept of norm of a matrix, then

$$\left|I - Q^{-1} \cdot A\right| < 1$$

Numerical algebra assures that if the equation above is fulfilled, then the system will converge to the solution. Note that the above expression will be smaller as Q^{-1} approximates A^{-1}. It is important to talk about the last part also, where we said "norm of a matrix." We know that vectors have a norm, which is their length, but this is our geometrical interpretation of them. Remember that vectors are abstract, therefore, they don't have a length, and similarly matrices also have a norm. Geometrical interpretations to this norm are abstract, but they can be viewed as how big the matrix is.

Usually many authors use an equivalent norm from vectors to matrices which is the Frobenius norm. The Frobenius norm is the square root of the sum of the squares of every element in the matrix. That is,

$$|A| = \sqrt{\sum_{i=1}^{n}\sum_{j=1}^{n} a_{ij}^2}$$

This field of algebra has more in it than just this, but this is the principle in which numerical methods rely on. All the others that come after these are developments of other authors, choices of Q, stability, round up errors, minimizing errors. It is truly a beautiful subject, but sadly it is outside the main theme of this book. Now that we have mentioned these terms and theoretical basis, let us study the procedures. Each procedure will depend on the authors and the choice of matrix Q.

Richardson's Method

Richardson's method is an iterative method that chooses matrix $\boldsymbol{Q}$ as the identity. Therefore, our main equation ends like this

$$\overrightarrow{x_n} = (I - A)^n \overrightarrow{x_0} - A^{-1}\left[(I - A)^n - I\right]\vec{b}$$

and the condition we need to assure convergence is this:

$$|I - A| < 1$$

Not many matrices can fulfill that condition. For this reason, Richardson's method works best in systems that involve matrices with small coefficients, similar to Markov matrices. Markov matrices are not included here because, if you look closely at the equation above, we could view as if all the eigenvalues of $I - A$ would have module less than 1 to assure convergence. But as we have seen in the previous chapters, Markov matrices have one eigenvalue equal to 1. Richardson's method is not used extensively nowadays, since there are better approaches and methods to real problems. However, it was developed at first because of its simplicity.

Jacobi's Method

Jacobi's method is another iterative method that chooses matrix $\boldsymbol{Q}$ as a diagonal matrix with the same entry elements as matrix $\boldsymbol{A}$. Therefore, when we compute $\left|I - Q^{-1}A\right|$ the result will be a new matrix whose diagonal elements will be all zero, also known as a hollow matrix (you should know why). This is an improvement to Richardson's method.

Gauss Seidel Method

Gauss Seidel method takes ***Q*** as a matrix which is the sum of a diagonal matrix and strictly lower triangular matrix which have the same coefficients as matrix ***A***. We could see it as if Gauss Seidel method splits ***A*** into three matrices, a diagonal matrix usually called ***D***, a strictly lower triangular matrix usually called ***L***, and a strictly upper triangular matrix usually called ***U***. Matrix ***Q*** is equal to the sum of the diagonal matrix and the lower triangular matrix. This is far greater than Jacobi's method since in the Gauss Seidel method, ***Q*** is more similar to ***A***, therefore, the inverse of ***Q*** is more similar to the inverse of ***A***.

$$A = \begin{pmatrix} a_{11} & a_{12} & \cdots & a_{1n} \\ a_{21} & a_{22} & \cdots & a_{2n} \\ \vdots & \vdots & \cdots & \vdots \\ a_{n1} & a_{n2} & \cdots & a_{nn} \end{pmatrix} = L + D + U$$

$$L = \begin{pmatrix} 0 & 0 & \cdots & 0 \\ a_{21} & 0 & \cdots & 0 \\ \vdots & \vdots & \cdots & \vdots \\ a_{n1} & a_{n2} & \cdots & 0 \end{pmatrix} D = \begin{pmatrix} a_{11} & 0 & \cdots & 0 \\ 0 & a_{22} & \cdots & 0 \\ \vdots & \vdots & \cdots & \vdots \\ 0 & 0 & \cdots & a_{nn} \end{pmatrix} U = \begin{pmatrix} 0 & a_{12} & \cdots & a_{1n} \\ 0 & 0 & \cdots & a_{2n} \\ \vdots & \vdots & \cdots & \vdots \\ 0 & 0 & \cdots & 0 \end{pmatrix}$$

$$A\vec{x} = \vec{b}$$

$$(L + D + U)\vec{x} = \vec{b}$$

$$\vec{x} = (L + D)^{-1} \cdot \left[\vec{b} - U \cdot \vec{x}\right]$$

Successive Over-Relaxation

The last numerical method that we are going to see is the Successive Over-Relaxation method, which is a generalization of all the methods we have mentioned. It is very close to Gauss Seidel but has its main differences. The successive over-relaxation method introduces a factor ù which is called the relaxation factor and must be between 0 and 2. This factor improves rate of convergence depending on its value. The formulation of the successive over-relaxation method is shown as follows.

$$A\vec{x} = \vec{b}$$

$$(L + D + U)\vec{x} = \vec{b}$$

$$(\omega L + (\omega - 1)D + \omega U + D)\vec{x} = \omega\vec{b}$$

$$(\omega L + D)\vec{x} = \omega\vec{b} - [(\omega - 1)D + \omega U]\vec{x}$$

$$\vec{x} = (\text{ù} L + D)^{-1} \cdot \left[\omega\vec{b} - [(\omega - 1)D + \omega U]\vec{x}\right]$$

If we take a closer look, the successive over relaxation method is the Gauss Seidel method for omega equal to 1. Let us discuss briefly about the relaxation factor. We said that it had to be between 0 and 2. However, this relaxation factor has to do with how fast the method converges or diverges. For example, it is shown that if omega is less than 1, then the solution will converge slower than Gauss Seidel; but if omega is more than 1, then the solution will converge faster. Thus, it is better to work with an omega higher than 1. However, it does not follow that the higher the omega is, that better it would be. It will still depend on the system, and the best choice of omega will probably come up after several simulations since a good choice of omega will make the problem behave better.

Now that we have seen this, let us have an example to see how these methods work. Suppose we have the following system of equations and we want to solve it by numerical methods:

$$\begin{pmatrix} 1 & 2 & 3 & 4 \\ 1 & 3 & 4 & 6 \\ 7 & 8 & 4 & 6 \\ 9 & 7 & 10 & 0 \end{pmatrix} \begin{pmatrix} x \\ y \\ z \\ w \end{pmatrix} = \begin{pmatrix} 6 \\ 8 \\ 15 \\ 11 \end{pmatrix}$$

Let us try with the first method which is ***Richardson's method.*** For this we need to compute $I - A$ and see if it satisfies the condition of $|I - A| < 1$. Therefore,

$$A = \begin{pmatrix} 1 & 2 & 3 & 4 \\ 1 & 3 & 4 & 6 \\ 7 & 8 & 4 & 6 \\ 9 & 7 & 10 & 0 \end{pmatrix} \rightarrow (I - A) = \begin{pmatrix} 0 & -2 & -3 & -4 \\ -1 & -2 & -4 & -6 \\ -7 & -8 & -3 & -6 \\ -9 & -7 & -10 & 1 \end{pmatrix}$$

$$|I - A| = 21,79$$

We can conclude that this is unlikely to converge with this method. We said "unlikely to converge" because we cannot conclude that this method will not work until we try. That is the thing with numerical methods -- we will never know what will happen unless we try it. However, this does not depend on luck. We have a well based theoretical field and every method has its restrictions and recommendations. Therefore, we can rest assured that Richardson's method **will** work if the norm of $I - A$ is less than one. However, we cannot assure that it will not converge if the norm of $I - A$ is greater than 1. Under these considerations, let us apply Richardson's method and look for convergence:

$$\overrightarrow{x_k} = (I - A)\overrightarrow{x_{k-1}} + \vec{b}$$

Now we only need a seed vector $\boldsymbol{x_0}$. Notice that if the initial vector we take is close to the solution, the method is likely to converge faster. As initial vector, let us choose the zero vector, and see what happens:

$$\overrightarrow{x_1} = (I - A)\overrightarrow{x_0} + \vec{b}$$

$$\overrightarrow{x_1} = \begin{pmatrix} 0 & -2 & -3 & -4 \\ -1 & -2 & -4 & -6 \\ -7 & -8 & -3 & -6 \\ -9 & -7 & -10 & 1 \end{pmatrix}\begin{pmatrix} 0 \\ 0 \\ 0 \\ 0 \end{pmatrix} + \begin{pmatrix} 6 \\ 8 \\ 15 \\ 11 \end{pmatrix} = \begin{pmatrix} 6 \\ 8 \\ 15 \\ 11 \end{pmatrix}$$

Now let us continue doing iterations and look for convergence:

$$\vec{x_2} = \begin{pmatrix} -99 \\ -140 \\ -202 \\ -238 \end{pmatrix}; \vec{x_3} = \begin{pmatrix} 1844 \\ 2623 \\ 3862 \\ 3664 \end{pmatrix} \text{...Divergency}$$

We can see that the solution for this method diverges greatly. However, this is not a reason to give up. Each method has great advantages over the others. This will also depend on the type of matrices we are working with. For example, some numerical methods work better with matrices with special properties like positive definite matrices, orthogonal matrices, diagonal matrices, triangular ones, symmetric matrices, and mostly all with diagonally dominant matrices. Let us look at how this problem works with ***Jacobi's method.*** Applying this method, we have

$$A = \begin{pmatrix} 1 & 2 & 3 & 4 \\ 1 & 3 & 4 & 6 \\ 7 & 8 & 4 & 6 \\ 9 & 7 & 10 & 0 \end{pmatrix}; Q = \begin{pmatrix} 1 & 0 & 0 & 0 \\ 0 & 3 & 0 & 0 \\ 0 & 0 & 4 & 0 \\ 0 & 0 & 0 & 0 \end{pmatrix}$$

Notice that we will encounter a problem trying to compute the inverse of matrix ***Q***. However, all is not lost. We can change our problem by exchanging some of the columns or rows of matrix A. Obviously, this will change our initial problem but that can be solved by exchanging variables or permute some coefficients on the independent term of the system. Let us suppose we decide to exchange two columns of matrix ***A***. Let us suppose they are the third and the fourth columns:

$$\begin{pmatrix} 1 & 2 & 4 & 3 \\ 1 & 3 & 6 & 4 \\ 7 & 8 & 6 & 4 \\ 9 & 7 & 0 & 10 \end{pmatrix} \begin{pmatrix} x \\ y \\ w \\ z \end{pmatrix} = \begin{pmatrix} 6 \\ 8 \\ 15 \\ 11 \end{pmatrix}$$

We noticed that the columns changed but also the order of our variables. Exchanging a row does a similar change. Let us suppose we want to change the fourth and second row of our original problem, we would get something like this.

$$\begin{pmatrix} 1 & 2 & 3 & 4 \\ 9 & 7 & 10 & 0 \\ 7 & 8 & 4 & 6 \\ 1 & 3 & 4 & 6 \end{pmatrix} \begin{pmatrix} x \\ y \\ z \\ w \end{pmatrix} = \begin{pmatrix} 6 \\ 11 \\ 15 \\ 8 \end{pmatrix}$$

This last step is doing a row operation. However, these manipulations are useful and sometimes necessary to solve real problems. From both of these approaches we can proceed with Jacobi's method, since our diagonal matrix $\boldsymbol{Q}$ will be invertible now. Let us stay with the column exchange. Therefore, our matrices would be

$$A = \begin{pmatrix} 1 & 2 & 4 & 3 \\ 1 & 3 & 6 & 4 \\ 7 & 8 & 6 & 4 \\ 9 & 7 & 0 & 10 \end{pmatrix}; Q = \begin{pmatrix} 1 & 0 & 0 & 0 \\ 0 & 3 & 0 & 0 \\ 0 & 0 & 6 & 0 \\ 0 & 0 & 0 & 10 \end{pmatrix}$$

And our iterations would be of the form

$$\overrightarrow{x_k} = \left(I - Q^{-1}A\right)\overrightarrow{x_k} + Q^{-1}\vec{b}$$

Let us compute all of these

$$\left(I - Q^{-1}A\right) = \frac{1}{3}\begin{pmatrix} 0 & -6 & -12 & -9 \\ -1 & 0 & -6 & -4 \\ -3.5 & -4 & 0 & -2 \\ -2.7 & -2.1 & 0 & 0 \end{pmatrix} Q^{-1} \cdot \vec{b} = \frac{1}{3}\begin{pmatrix} 18 \\ 11 \\ 7.5 \\ 2.4 \end{pmatrix}$$

where all diagonal entries are zero. Let us calculate the norm of $I - Q^{-1}A$ to see if we can assure convergence.

$$\left|I - Q^{-1}A\right| = 6.3065$$

Thus, we cannot assure convergence. However, this norm is closer to 1 than the one we had on Richardson's method. Let us iterate and see if we get convergence. Let us also start with the zero vector as seed.

$$\vec{x}_1 = \begin{pmatrix} 6 \\ 3.66 \\ 2.5 \\ 0.8 \end{pmatrix}; \vec{x}_2 = \begin{pmatrix} -13.733 \\ -4.4 \\ -9.92 \\ -7.16 \end{pmatrix}; \vec{x}_3 = \begin{pmatrix} 75.98 \\ 37.64 \\ 29.16 \\ 16.24 \end{pmatrix} ...Divergency(again)$$

We had a divergent solution again, so there should be something that we are not doing quite right. But we are doing all procedures perfectly right. However, we must be clever when working with matrices. If you notice from our last method, we need the norm of the matrix we are working to be less than 1. Let us discuss a case that will surely guaranty convergences. If our matrix would be diagonally dominant then we can assure convergence with Jacobi's method. We will see why in a moment, but before that, let us take a look at our original system:

$$\begin{pmatrix} 1 & 2 & 3 & 4 \\ 1 & 3 & 4 & 6 \\ 7 & 8 & 4 & 6 \\ 9 & 7 & 10 & 0 \end{pmatrix} \begin{pmatrix} x \\ y \\ z \\ w \end{pmatrix} = \begin{pmatrix} 6 \\ 8 \\ 15 \\ 11 \end{pmatrix}$$

Let us try by a set of row and column exchanges to get the matrix diagonally dominant, like this:

$$\begin{pmatrix} 1 & 4 & 2 & 3 \\ 1 & 6 & 3 & 4 \\ 7 & 6 & 8 & 4 \\ 9 & 0 & 7 & 10 \end{pmatrix} \begin{pmatrix} x \\ w \\ y \\ z \end{pmatrix} = \begin{pmatrix} 6 \\ 8 \\ 15 \\ 11 \end{pmatrix}$$

where three out of four rows of the matrix are diagonally dominant, which may be enough to get convergence. If we would apply Jacobi's method using the same procedure as we did before,

$$A = \begin{pmatrix} 1 & 4 & 2 & 3 \\ 1 & 6 & 3 & 4 \\ 7 & 6 & 8 & 4 \\ 9 & 0 & 7 & 10 \end{pmatrix}; Q = \begin{pmatrix} 1 & 0 & 0 & 0 \\ 0 & 6 & 0 & 0 \\ 0 & 0 & 8 & 0 \\ 0 & 0 & 0 & 10 \end{pmatrix}$$

$$I - Q^{-1}A = \frac{1}{3}\begin{pmatrix} 0 & -6 & -12 & -9 \\ -1 & 0 & -6 & -4 \\ -3.5 & -4 & 0 & -2 \\ -2.7 & -2.1 & 0 & 0 \end{pmatrix} Q^{-1}\cdot\vec{b} = \frac{1}{3}\begin{pmatrix} 18 \\ 4 \\ 5.625 \\ 3.3 \end{pmatrix}$$

If we start iterating with these new elements, choosing as seed the zero vector,

$$\vec{x_1}=\begin{pmatrix}6\\1.33\\1.875\\1.1\end{pmatrix};\vec{x_2}=\begin{pmatrix}-6.38\\-1.33\\-4.92\\-5.61\end{pmatrix};\vec{x_3}=\begin{pmatrix}38.03\\8.60\\11.26\\10.29\end{pmatrix}...Divergency(thirdtime)$$

Notice that did not get convergence at all, but we diverged slower than before (check at the values we got with respect to other iterations). Thus, we are conditioning the system for a better simulation. Let us try with ***Gauss Seidel*** method. Gauss Seidel method is much developed than the ones we have studied so far, so we might have a chance for it to converge. Let us continue with our problem and take a look at the last matrix we computed.

$$A=\begin{pmatrix}1&4&2&3\\1&6&3&4\\7&6&8&4\\9&0&7&10\end{pmatrix};Q=\begin{pmatrix}1&0&0&0\\1&6&0&0\\7&6&8&0\\9&0&7&10\end{pmatrix}$$

Let us compute the necessary elements.

$$I-Q^{-1}A=\frac{1}{3}\begin{pmatrix}0&-12&-6&-9\\0&2&-0.5&-0.5\\0&9&5.62&6.75\\0&4.5&1.46&3.37\end{pmatrix}\quad Q^{-1}\vec{b}=\frac{1}{3}\begin{pmatrix}18\\1\\-10.87\\-5.28\end{pmatrix}\quad \left|I-Q^{-1}A\right|=7.13$$

Iterating,

$$\vec{x_1}=\begin{pmatrix}6\\0.33\\-3.62\\-1.76\end{pmatrix};\vec{x_2}=\begin{pmatrix}17.20\\1.45\\-13.38\\-5.01\end{pmatrix};\vec{x_3}=\begin{pmatrix}41.99\\4.36\\-35.64\\-11.74\end{pmatrix}...Divergency(fourthtime)$$

This problem is surely a tough one to solve with numerical methods. But to finish this section, let us finalize with the ***successive over-relaxation*** method which is a generalization of all. For this method we need to split A in the following way:

$$A=\begin{pmatrix}1&4&2&3\\1&6&3&4\\7&6&8&4\\9&0&7&10\end{pmatrix};L=\begin{pmatrix}0&0&0&0\\1&0&0&0\\7&6&0&0\\9&0&7&0\end{pmatrix};D=\begin{pmatrix}1&0&0&0\\0&6&0&0\\0&0&8&0\\0&0&0&10\end{pmatrix};U=\begin{pmatrix}0&4&2&3\\0&0&3&4\\0&0&0&4\\0&0&0&0\end{pmatrix}$$

Let us choose for this example, omega equal to 1.3 and compute the results:

$$(\omega L+D)^{-1}\cdot\left[(\omega-1)D+\omega U\right]=\begin{pmatrix}0.3&5.2&2.6&3.9\\-0.065&-0.82&0.086&0.021\\-0.277&-5.10&-2.74&-3.80\\-0.098&-1.435&-0.546&-0.8\end{pmatrix}$$

$$\omega(\omega L+D)^{-1}\cdot\vec{b}=\begin{pmatrix}7.8\\0.0433\\-6.477\\-1.801\end{pmatrix}$$

Let us start to iterate, hoping for convergence this time:

$$\vec{x_1}=\begin{pmatrix}7.8\\0.0433\\-6.477\\-1.801\end{pmatrix};\vec{x_2}=\begin{pmatrix}29.10\\1.186\\-28.70\\-5.95\end{pmatrix};\vec{x_3}=\begin{pmatrix}90.76\\5.53\\-93.71\\-17.69\end{pmatrix}...Divergency(lasttime)$$

As you can see, this problem cannot be solved by any numerical method that we have discussed. However, we learned several things: that numerical methods need some conditions to achieve convergence; sometimes they are not the best way to approach a problem, and; they work better if we make rearrangements of the initial system. As a handyman, we need to use the right tool for the right job. An advice with numerical methods is that they work perfectly if matrices are diagonally dominant, or at least positive definite matrices.

Now that we showed numerical methods for this problem, we should approach it by Gaussian elimination.

$$\begin{pmatrix}x\\y\\z\\w\end{pmatrix}=\begin{pmatrix}2\\3/2\\-1\\0\end{pmatrix}$$

As a closing remark, let us emphasize again, that we just scratched the surface of an enormously large and well-developed field of numerical mathematics.

COMPUTING EIGENVALUES AND EIGENVECTORS

In the past chapter, we saw what eigenvalues and eigenvectors are, however, have not seen how to calculate them on high size matrices. There are many methods to do this, but let us take a look first at ***Gershgorin circle theorem***. Let us use one of the examples that we had before.

$$A = \begin{pmatrix} 5 & 4 & 2 & 1 \\ 0 & 1 & -1 & -1 \\ -1 & -1 & 3 & 0 \\ 1 & 1 & -1 & 2 \end{pmatrix}$$

From procedures we did before, we know that this matrix has eigenvalues equal 1, 2 and 4. Assuming we do not know this, Gershgorin theorem gives us the exact region where they might be. According to the theorem, we will have 1 disc for each row, and they will have center equal to the element in the diagonal and radius equal to the sum of the absolute values of the other elements in that row. Therefore, our four discs are the following:

$$D_i = \left(a_{ii}, \sum_{j \neq i} \left| a_{ij} \right| \right) \rightarrow D_1(5,7); D_2(1,2); D_3(3,2); D_4(2,3)$$

If we graph this regions to get a graphical view of where our eigenvalues are, we get:

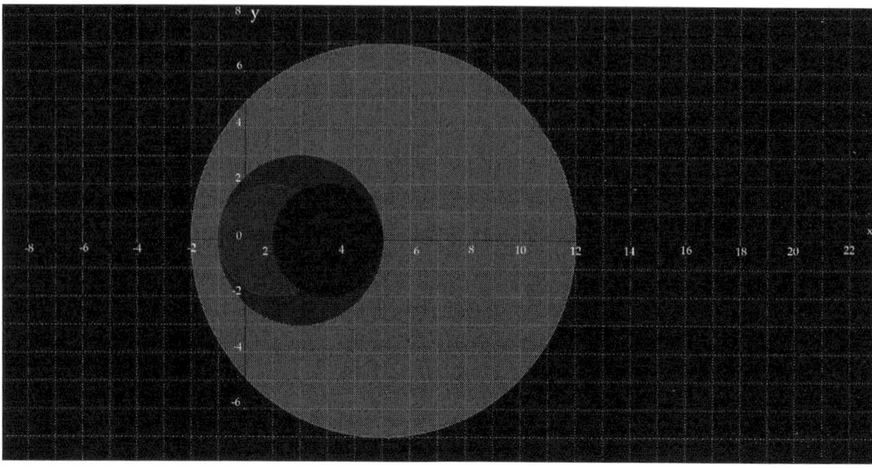

Figure 16: Location of eigenvalues based on Gershgorin theorem.

We can see in the figures above where our eigenvalues reside, that is, on the four circles. Note that theses circles also include any possible complex eigenvalue. You can also see that the circles overlap. There are two important things to remember about Gershgorin circles.

One of them is that every circle is not required to contain an eigenvalue, therefore, we are working with the union of all circles. And as all overlap, we can say that the eigenvalues are within the following disc $D(5, 7)$. It is true that every circle does not necessarily contain an eigenvalue, however, the fact that the theorem reduces the possibilities is very much appreciated.

The other detail is that Gershgorin circles also work with the sum of the absolute values of the columns (which would be the proof of the theorem but with A^T), because ***the eigenvalues of a matrix are also the eigenvalues of its transpose***. If we do this approach with the columns of matrix ***A***, we get:

$$D_j = \left(a_{jj}, \sum_{i \neq j} |a_{ij}| \right) \rightarrow D_1(5,2); D_2(1,6); D_3(3,4); D_4(2,2)$$

Graphing this, we get

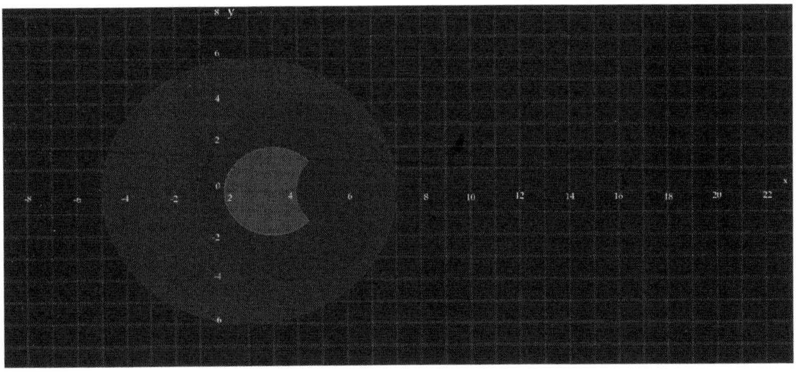

Figure 17: Application of the corollary of the Gershgorin theorem.

Which is a new region. Again, as all circles overlap, we can say that all the eigenvalues are contained within a disk of the form $D(1,6)$. But it does not end here. The fact that we found two regions in which all the eigenvalues live means that the eigenvalues must be in the intersection of both regions (to satisfy all the conditions). Therefore, we can reduce the problem even more intersecting both regions. This is highly appreciated while looking for eigenvalues numerically because we have a region where we can find our eigenvalues.

NUMERICAL METHODS TO COMPUTE EIGENVALUES AND EIGENVECTORS

There are two methods that have been developed and used for the past five decades that are standard now to calculate eigenvalues and eigenvectors for moderate size matrices. From a mathematical and engineering point of view, these are methods that are relatively new, but they are still being used for their reliability and easement.

The first method is ***the power method***, which consists of taking a seed vector and multiplying it over and over again by the powers of a matrix $\boldsymbol{A}$ – the matrix from which we want the eigenvalues and eigenvectors. However, this method only gives the greatest eigenvalue and its corresponding eigenvector. The proof is rather simple, suppose we have a vector $\boldsymbol{x}_0$ which can be decomposed as a linear combination of the eigenvectors of our matrix $\boldsymbol{A}$. Let us write this:

$$x_0 = c_1\vec{e_1} + c_2\vec{e_2} + \ldots + c_n\vec{e_n}$$

If we multiply by A^k at both sides, we get:

$$A^k x_0 = A^k\left[c_1\vec{e_1} + c_2\vec{e_2} + \ldots + c_n\vec{e_n}\right]$$

We can transform this into:

$$A^k x_0 = c_1\lambda_1^k\vec{e_1} + c_2\lambda_2^k\vec{e_2} + \ldots + c_n\lambda_n^k\vec{e_n}$$

If we extract common factor the greatest eigenvalue of A, let us assume is $\boldsymbol{\lambda_1}$, we get:

$$A^k x_0 = \lambda_1^k\left[c_1\vec{e_1} + c_2\left(\frac{\lambda_2}{\lambda_1}\right)^k\vec{e_2} + \ldots + c_n\left(\frac{\lambda_n}{\lambda_1}\right)^k\vec{e_n}\right]$$

As ***k*** goes to infinity, all the lambda quotients approximate zero, therefore,

$$\lim_{k\to\infty} A^k x_0 = \lambda_1^k c_1\vec{e_1}$$

As you can see, we can get an eigenvector by doing this approach. To get the eigenvalue, some books take the dot product between the calculated eigenvector and other arbitrary vector ***w*** and take the next quotient:

$$\lambda_1 \approx \frac{\vec{w}\cdot A^{k+1}\overrightarrow{x_0}}{\vec{w}\cdot A^k\overrightarrow{x_0}}$$

which does converge to the eigenvalue. However, other authors calculate the eigenvalue like this:

$$\lambda = \frac{\overrightarrow{e_1^T}\cdot A\overrightarrow{e_1}}{\overrightarrow{e_1^T}\cdot\overrightarrow{e_1}}.$$

The next method we are going to see is ***the QR method***, which corresponds to transform our matrix ***A*** into a ***QR*** form, then flipping the order of both matrices, obtaining a new matrix that will be similar to ***A*** and repeating the process. This algorithm assures that after a certain number of iterations we will get a matrix that will be an upper triangular matrix. Thus, we will be able to get its eigenvalues straightforward. To see this mathematically, the ***QR*** algorithm consists on the following:

$$A_0 = Q_0 \cdot R_0 \rightarrow A_1 = R_0 \cdot Q_0 \rightarrow A_1 = Q_1 \cdot R_1 \rightarrow \cdots$$

This method relies in the fact that this process transforms A into a similar matrix, a different matrix but with the same eigenvalues, because of this:

$$A_{k+1} = R_k \cdot Q_k = Q_k^T \cdot Q_k \cdot R_k Q_k = Q_k^T A_k Q_k$$

There are other more sophisticated methods, improvements of the ones we just saw and special algorithms for special matrices (mainly sparse matrices), however, explaining those would be far too detailed for this book. It does not mean, however, that the methods we discussed are not suitable for tackling great problems. In fact, these methods are very easy to use, implement, fast, and above all, are an easier way to approach eigenvalue and eigenvector problems than calculating a determinant.

To show how these methods work, let us show an example. Suppose we have the same matrix as before:

$$A = \begin{pmatrix} 1 & 2 & 3 & 4 \\ 1 & 3 & 4 & 6 \\ 7 & 8 & 4 & 6 \\ 9 & 7 & 10 & 0 \end{pmatrix}$$

But now let us suppose that we want to calculate its eigenvalues. Let us do this by the methods we just saw. The first one of them was the power method which involves taking powers of $\boldsymbol{A}$ and multiplying them by a given vector. After a finite number of steps we would get an approximation to the eigenvector of $\boldsymbol{A}$ corresponding to the largest eigenvalue. Before doing this, let us use Gershgorin theorem to see where the eigenvalues might be. Doing this we get four circles for the rows of $\boldsymbol{A}$ and four circles for the columns of $\boldsymbol{A}$:

$$A = \begin{pmatrix} 1 & 2 & 3 & 4 \\ 1 & 3 & 4 & 6 \\ 7 & 8 & 4 & 6 \\ 9 & 7 & 10 & 0 \end{pmatrix}$$

$$Rowdiscs: D_1 = (1,9); D_2 = (3,11); D_3 = (4,21); D_4 = (0,26)$$

$$Column\,discs: D_1 = (1,17); D_2 = (3,17); D_3 = (4,17); D_4 = (0,16)$$

If we graph both regions and intersect them we get:

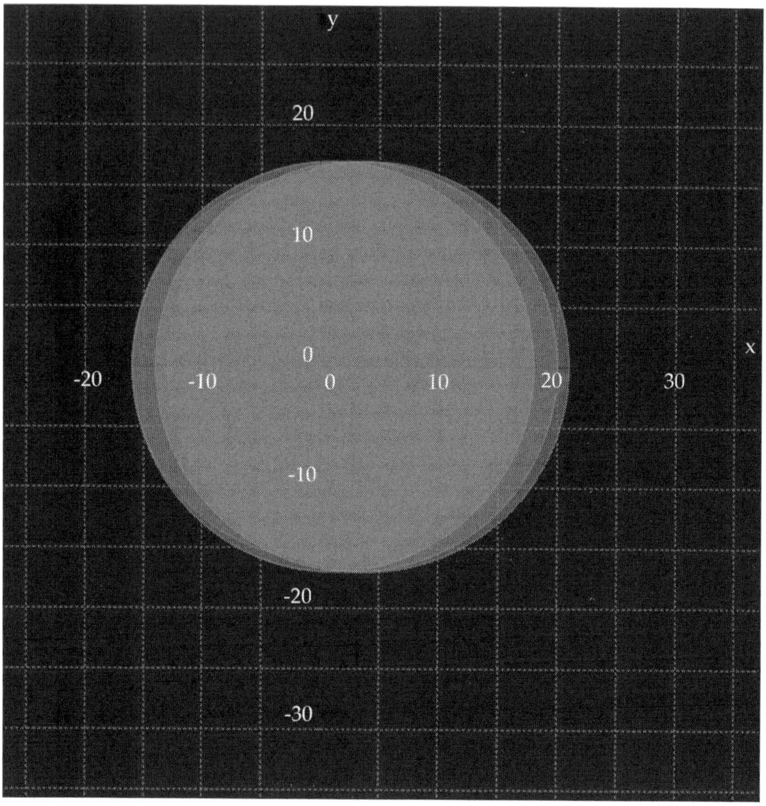

Figure 18: Disks containing the eigenvalues.

That all the eigenvalues (which might be real or complex) are somewhere between -16 and 21. To find the highest eigenvalue we proceed to take powers of ***A*** and multiply by an arbitrary vector, let us call it vector ***u***. Let us assume vector ***u*** is of the following and start iterating:

$$\vec{v_0} = \vec{u} = \begin{pmatrix} 1 \\ 1 \\ 1 \\ 1 \end{pmatrix}; \vec{v_1} = A \cdot \vec{u} = \begin{pmatrix} 10 \\ 14 \\ 25 \\ 26 \end{pmatrix}; \vec{v_2} = A^2 \cdot \vec{u} = \begin{pmatrix} 217 \\ 308 \\ 438 \\ 438 \end{pmatrix}; \vec{v_3} = A^3 \cdot \vec{u} = \begin{pmatrix} 3899 \\ 5521 \\ 8363 \\ 8489 \end{pmatrix}; \vec{v_4} = A^4 \cdot \vec{u} = \begin{pmatrix} 73986 \\ 104848 \\ 155847 \\ 157368 \end{pmatrix}$$

$$\overrightarrow{v_5} = A^5 \cdot \vec{u} = \begin{pmatrix} 1380695 \\ 1956126 \\ 2924282 \\ 2958280 \end{pmatrix}$$

To calculate the eigenvalue, two approaches can be done. The first one is to multiply the results we just got by an arbitrary vector, let us call it ***w***, and divide them by the last known product, in other words, this:

$$\lambda_n = \frac{\vec{w} \cdot \overrightarrow{v_n}}{\vec{w} \cdot \overrightarrow{v_{n-1}}}$$

Assuming the easiest possibility of vector ***w***, we get:

$$\vec{w} = \begin{pmatrix} 1 \\ 0 \\ 0 \\ 0 \end{pmatrix} \rightarrow \lambda_1 = \frac{\vec{w} \cdot \overrightarrow{v_1}}{\vec{w} \cdot \overrightarrow{v_0}} = 10; \lambda_2 = 21.7; \lambda_3 = 17.96; \lambda_4 = 18.97; \lambda_5 = 18.66; \lambda_6 = 18.75$$

Another approach to look for the eigenvalue is to start from the equation:

$A\vec{v} = \lambda \vec{v}$

Multiply at both sides by the transpose of vector ***v*** and solve for lambda:

$$\lambda = \frac{\overrightarrow{v^T} \cdot A\vec{v}}{\overrightarrow{v^T} \cdot \vec{v}}$$

Doing this approach we get:

$\lambda_1 = 18.04; \lambda_2 = 18.886; \lambda_3 = 18.6831; \lambda_4 = 18.75; \lambda_5 = 18.73; \lambda_6 = 18.73$

Which we see is a faster approach. In order to obtain the eigenvector what we need to do is normalize the vectors ***v*** we got, doing this we get that

$$\widehat{v_1} = \begin{pmatrix} 0.2502 \\ 0.3503 \\ 0.6255 \\ 0.6506 \end{pmatrix}; \widehat{v_2} = \begin{pmatrix} 0.2993 \\ 0.4248 \\ 0.6041 \\ 0.6041 \end{pmatrix}; \widehat{v_3} = \begin{pmatrix} 0.2846 \\ 0.4029 \\ 0.6104 \\ 0.6196 \end{pmatrix}; \widehat{v_4} = \begin{pmatrix} 0.2890 \\ 0.4096 \\ 0.6088 \\ 0.6148 \end{pmatrix}; \widehat{v_5} = \begin{pmatrix} 0.2876 \\ 0.4075 \\ 0.6092 \\ 0.6163 \end{pmatrix}; \widehat{v_5} = \begin{pmatrix} 0.2881 \\ 0.4082 \\ 0.6091 \\ 0.6158 \end{pmatrix}$$

From where we see that we got the largest eigenvalue and its corresponding eigenvector with two known significant figures, in only 6 iterations, but now we need to find the other 3 eigenvalues. The way to do this with the power method is to look at ***two properties of the eigenvalues***.

The first one is that the eigenvalues of the inverse of ***A*** are also the inverses of the eigenvalues of ***A***. This can be proved from the main eigenvalue equation by multiplying by the inverse of matrix ***A*** and rearranging like this.

$$A\vec{v} = \lambda\vec{v}$$

$$A^{-1}A\vec{v} = A^{-1}\lambda\vec{v}$$

$$\vec{v} = \lambda A^{-1}\vec{v}$$

$$A^{-1}\vec{v} = \frac{1}{\ddot{e}}\vec{v}$$

As you can see, matrix ***A*** and its inverse have the same eigenvectors, and the eigenvalues of ***A*** are the inverses of the eigenvalues of A^{-1}.

The second property that we need to look at is the following. Suppose we have a matrix ***A***, but we shift its diagonal by a certain amount, let us say ***p***. Shifting the diagonal means doing this to a generic matrix ***A***:

$$A - pI = \begin{pmatrix} a_{11} - p & a_{12} & \cdots & a_{1n} \\ a_{21} & a_{22} - p & \cdots & a_{2n} \\ \vdots & \vdots & \cdots & \vdots \\ a_{n1} & a_{n2} & \cdots & a_{nn} - p \end{pmatrix}$$

And doing this also shifts the eigenvalues of ***A*** by ***p***. Meaning,

$$A\vec{v} = \lambda\vec{v}$$

$$(A - pI)\vec{v} = (\lambda - p)\vec{v}$$

This preserves the eigenvectors, too. Let us look at these two properties and see how to combine them with the power method to achieve our goal.

Suppose we have a matrix ***A*** of size ***n*** with ***n*** eigenvalues, and they are listed from greatest to smallest in the following way:

$$A\vec{v} = \lambda\vec{v} \rightarrow \lambda_1 \geq \lambda_2 \geq \lambda_3 \geq \ldots \geq \lambda_n$$

From the past two properties we know we can rearrange the above expression to get the eigenvalues in the following way:

$$(A - pI)^{-1}\vec{v} = \frac{1}{\lambda - p}\vec{v} \rightarrow \frac{1}{\lambda_p - p} \geq \frac{1}{\lambda_{p-1} - p} \geq \ldots \frac{1}{\lambda_{p-n} - p}$$

Where the different λ_p are the eigenvalues that are closer to point ***p***. When the eigenvalues get closer to point ***p,*** their subtraction gets smaller, therefore their multiplicative inverse gets bigger. This allows us to apply the power method with this new conditions to find the eigenvalues that we are looking for. However, we must be careful because we should not input a value of ***p*** very close to an eigenvalue because $A - pI$ would be very difficult to invert (and if ***p*** would be an eigenvalue, then we could not invert that matrix). The only inconvenience with this method is that it is time consuming and needs to calculate the inverse of a matrix. However, many improvements can be done to it, like the Arnoldi method, but it is numerically unstable and we will not cover it in this book.

Going back to our main problem, let us choose ***p*** equal to zero and apply the power method. This will allow us to calculate the smallest eigenvalue of ***A***. If we iterate with the same vector ***u***, we get:

$$A = \begin{pmatrix} 1 & 2 & 3 & 4 \\ 1 & 3 & 4 & 6 \\ 7 & 8 & 4 & 6 \\ 9 & 7 & 10 & 0 \end{pmatrix} A^{-1} = \begin{pmatrix} 3.1915 & -2.2765 & 0.1489 & -0.1063 \\ -3.8297 & 2.5319 & 0.0212 & 0.1276 \\ -0.1914 & 0.2766 & -0.1489 & 0.1063 \\ 1.5106 & -0.9042 & 0.0638 & -0.1170 \end{pmatrix}$$

$$\vec{v_0} = \vec{u} = \begin{pmatrix} 1 \\ 1 \\ 1 \\ 1 \end{pmatrix}; \vec{v_1} = A^{-1} \cdot \vec{u} = \begin{pmatrix} 0.9574 \\ -1.1489 \\ 0.0425 \\ 0.5532 \end{pmatrix}; \vec{v_2} = \begin{pmatrix} 5.6188 \\ -6.5043 \\ -0.4486 \\ 2.4232 \end{pmatrix}; \vec{v_3} = \begin{pmatrix} 32.4154 \\ -37.6874 \\ -2.5504 \\ 14.0573 \end{pmatrix}; \vec{v_4} = \begin{pmatrix} 187.37 \\ -217.82 \\ -14.7561 \\ 81.2393 \end{pmatrix}$$

$$\lambda_1^{-1} = 5.5703; \lambda_2^{-1} = 5.7845; \lambda_3^{-1} = 5.7800; \lambda_4^{-1} = 5.7801$$

$$\widehat{v_1} = \begin{pmatrix} 0.6002 \\ -0.7202 \\ 0.0266 \\ 0.3467 \end{pmatrix}; \widehat{v_2} = \begin{pmatrix} 0.6284 \\ -0.7274 \\ -0.0501 \\ 0.2710 \end{pmatrix}; \widehat{v_3} = \begin{pmatrix} 0.6267 \\ -0.7286 \\ -0.0493 \\ 0.2717 \end{pmatrix}; \widehat{v_4} = \begin{pmatrix} 0.6267 \\ -0.7286 \\ -0.0493 \\ 0.2717 \end{pmatrix}$$

Let us remember that the resulting eigenvector is an eigenvector of A, but this cigenvalue is the inverse of the closer eigenvalue of A to zero, so we must invert our final result. Let us show our results so far:

$$\lambda_1 = 18.73 \to \hat{e_1} = \begin{pmatrix} 0.2881 \\ 0.4082 \\ 0.6091 \\ 0.6158 \end{pmatrix} \lambda_2 = \frac{1}{\lambda_4^{-1}} = 0.1730 \to \hat{e_2} = \begin{pmatrix} 0.6267 \\ -0.7286 \\ -0.0493 \\ 0.2717 \end{pmatrix}$$

To calculate the other eigenvalues of A we will use the fact that the sum of all the eigenvalues is equal to the trace and the product of all the eigenvalues is equal to the determinant of A. Calculating the determinant of A in this case is easy because we count with elemental operations to reduce the matrix and solve for its determinant. By doing this we will get

$$\lambda_1 + \lambda_2 + \lambda_3 + \lambda_4 = tr(A) = 8$$

$$\lambda_1 \cdot \lambda_2 \cdot \lambda_3 \cdot \lambda_4 = det(A) = 94$$

Solving for the remaining eigenvalues we get

$$\lambda_1 = 18.73; \lambda_2 = 0.1730; \lambda_3 = -6.2936; \lambda_4 = -4.6094$$

In order to get the remaining eigenvectors, two things can be done. One is to apply the power method with the inverse of A shifted very close to the eigenvalues, and the other one is to go back to the main equation and calculate the null space of $A - \lambda I$ ..

It is important to say that, in case we would use the power method with the inverse of A shifted, then p could not be an eigenvalue, it should be very close, but cannot be an eigenvalue, because if p is an eigenvalue then $(A - p \cdot I)^{-1}$ is not invertible.

- Between both methods, it is better to input the eigenvalue and calculate the null space, because this would be a lot easier than calculating an inverse and apply the power method again. However, it will be up to you.

Let us now use another method to calculate the eigenvalues of A, which is the ***QR method***. This same problem was done in chapter 4, where we found out that the ***QR*** decomposition is as follows:

$$A = Q \cdot R$$

$$Q^T = Q_3 \cdot Q_2 \cdot Q_1 = \begin{pmatrix} 0.0870 & 0.0870 & 0.6092 & 0.7833 \\ 0.3437 & 0.6680 & 0.4617 & -0.4715 \\ 0.3770 & 0.5328 & -0.6436 & 0.3995 \\ -0.8555 & 0.5121 & -0.0361 & 0.0662 \end{pmatrix}$$

$$R = Q_3 \cdot Q_2 \cdot Q_1 \cdot A = \begin{pmatrix} 11.489 & 10.792 & 10.879 & 4.526 \\ 0 & 3.0846 & 0.8349 & 8.1536 \\ 0 & 0 & 4.6831 & 0.8433 \\ 0 & 0 & 0 & -0.5666 \end{pmatrix}$$

According to the ***QR*** method, to get the eigenvalues of ***A***, we need to invert this order and get another matrix which is similar to ***A***:

$$A = Q \cdot R \to A_1 = R \cdot Q = \begin{pmatrix} 12.11 & 14.05 & 4.888 & -4.396 \\ 7.164 & -1.398 & 4.363 & 2.090 \\ 3.514 & 1.764 & -2.677 & -0.113 \\ -0.443 & 0.267 & -0.226 & -0.037 \end{pmatrix}$$

If we continue this method, we will get several matrices which are similar to ***A*** in the following order:

$$A_2 = \begin{pmatrix} -18.797 & 1.066 & -3.606 & 3.926 \\ -4.668 & 6.690 & -2.261 & 0.818 \\ 1.329 & -3.393 & -4.239 & 0.606 \\ -0.004 & 0.002 & -0.006 & 0.124 \end{pmatrix}$$

$$A_3 = \begin{pmatrix} -17.848 & 8.917 & -1.012 & -3.960 \\ 1.626 & 5.598 & 3.993 & 0.544 \\ 0.316 & 2.103 & -4.093 & -0.716 \\ 0 & 0 & 0 & 0.123 \end{pmatrix}$$

$$A_4 = \begin{pmatrix} -17.848 & 8.917 & -1.012 & -3.960 \\ 1.626 & 5.598 & 3.993 & 0.544 \\ 0.316 & 2.103 & -4.093 & -0.716 \\ 0 & 0 & 0 & 0.123 \end{pmatrix}$$

$$A_5 = \begin{pmatrix} -18.352 & 7.374 & -2.127 & -3.977 \\ 0.244 & 6.619 & 2.864 & 0.245 \\ 0.023 & 1.128 & -4.610 & -0.786 \\ 0 & 0 & 0 & 0.123 \end{pmatrix}$$

$$A_6 = \begin{pmatrix} -18.446 & 6.577 & 3.326 & 3.979 \\ -0.093 & 7.054 & -0.860 & -0.057 \\ -0.007 & 0.838 & -4.951 & -0.812 \\ 0 & 0 & 0 & 0.123 \end{pmatrix}$$

$$A_7 = \begin{pmatrix} -18.412 & 7.053 & -2.511 & -3.978 \\ 0.035 & 6.849 & 2.281 & 0.173 \\ 0.002 & 0.569 & -4.780 & -0.799 \\ 0 & 0 & 0 & 0.123 \end{pmatrix}$$

You can see that these coefficients look like the eigenvalues. However, they are somewhat different from the ones calculated from the power method. This happens because of taking too few decimal places. Besides, we still need to make some iterations to complete the method. Note that there convergence, which is a good sign.

CHAPTER 7

COMMON TOPICS IN ALGEBRA

"Calculus, the electrical battery, the telephone, the steam engine, the radio - all these groundbreaking innovations were hit upon by multiple inventors working in parallel with no knowledge of one another."

Steven Johnson

INTRODUCTION

For the past six chapters of this book, we have talked about several concepts that form a structure on which other areas of algebra rely on. These concepts form a very well based, profound and intriguing body of linear algebra. However, with the groundbreaking innovations that have happened since the nineteenth century, knowledge has been growing at an alarming rate. We may have noticed this, because one and a half centuries ago we do not even have electrical current in our houses, nor cars, nor television, nor computers, nor internet.

These advances do not happen on their own. They happen because science evolves with man and man evolves with science. In the last decade many authors have the same idea that humanity will probably always want to improve its surroundings, automate processes, increase its knowledge and have easy access to it. Nevertheless, having too much knowledge sometimes makes it difficult to focus, because sciences are getting so refined, so profound and so detailed that traveling through their seas becomes a long journey.

For this reason, this chapter is dedicated to bring special attention to important subjects that usually do not fit under the scope of a central body of knowledge.

SPARSE MATRICES

To start this chapter we will start talking about sparse matrices. ***Sparse matrices*** are matrices that own many zero entries in them. This kind of matrices appear in many places, mostly in engineering problems that are loosely coupled, that is, problems that do not have many restrictions on them. These kind of matrices represent a great opportunity for engineers, programmers and mathematicians because they have a special property -- they have many zeros in them!

Before getting deeper into this, let us think about sparsity. A matrix of a small size, like amatrix with zeros in it, can be hardly considered sparse, since there are not many elements in the matrix nor variables. Besides, a matrix of this size with some zeros does not represent a major change. But as sizes and variables grow in number, sparsity becomes a more desirable quality, like in a 50×50 matrix or in a 500×500 one, for obvious reasons.

The past idea has an immediate consequence. That, because of their great size, sparse matrices must be dealt with computers and not by a pencil and paper approach. From here came many opportunities for engineers and programmers to start an optimization process, beginning from the storage aspect. This has to do more with programming, however, if the reader would like to develop his own software it would be nice to give a hint on where to start. Storing a sparse matrix means to take advantage of its zeros, so several formats have been developed for such a task, like the Dictionary of Keys (DOK), the List of Lists (LIL), the Compressed Sparse Row (CSR) and the Compressed Sparse Column (CSC).

Probably the reader might ask himself how much of storage can be saved by using different formats of storage. To answer this question, let us look at a picture of a sparse matrix:

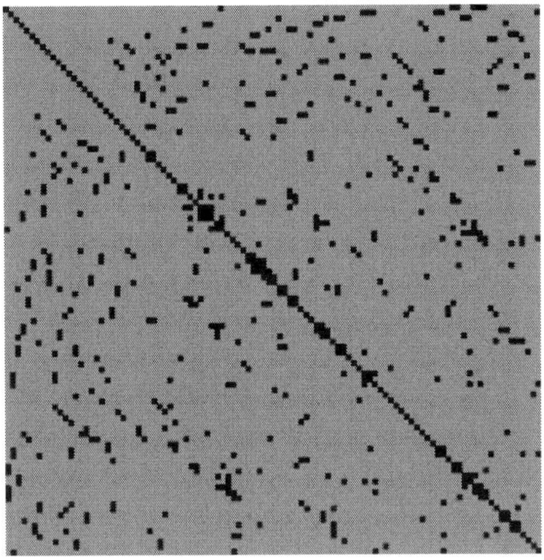

Figure 19: Sparse matrix.

Source: wikipedia.org

where each non-white square represents a nonzero coefficient. From the above image we can see that we have a high level of sparsity, since almost all of the elements in the matrix are zero. Actually, there is a way of measuring sparsity in a matrix, which is the number of zero elements divided by the number of elements in the matrix. A matrix is called sparse if most of its elements are zero and is called dense if most of its elements are nonzero.

In the preceding examples most matrices where dense, which means we had very well coupled systems but, in many cases, it is common and very well expected to obtain a sparse matrix while studying a certain phenomenon. This can be seen from the above image, which corresponds to a regular system of equations in a Finite Element Analysis (the above image also corresponds to a symmetric matrix). However, these situations also appear while making a discretization while solving partial differential equations. A final comment on this topic is that, usually, while taking smaller elements in this kinds of study of finite element analysis, the resulting matrix will be sparser but the solution will be more accurate.

Besides investing energy in how to store a sparse matrix, a lot of effort has also been implemented on how to solve systems with sparse matrices. For example, if we would solve a system of equations involving a sparse matrix it would not be pleasant to solve it using the inverse of the coefficient matrix, since a sparse matrix usually does not have a sparse inverse.

In reality calculating an inverse matrix is not a thing we would normally do, since it would take an immense quantity of calculations, prone to rounding errors, time consuming just to solve only a regular system of equations that may change in a couple of seconds by taking other initial conditions! This situation makes the user fall for numerical methods, because they offer a better way to get an easy solution at the cost of having to repeat the iterative method with each condition.

What is usually done when working with sparse systems is to factor the matrix into an LU form, or in case it is a diagonal dominant symmetric matrix, factor it using a Cholesky's decomposition, because it is assured that a sparse matrix will have a sparse decomposition (which is awesome to work with). However, the reader might have noticed that symmetric matrices share incredible properties, and in this case it is not an exception. In case a symmetric, positive definite sparse matrix appears while solving a system of equations, then usually what is done is to apply a numerical approach like the Steepest Descent method or the Conjugate Gradient method, which are numerical methods specially designed for symmetric positive definite matrices.

CONJUGATE GRADIENT METHOD

The ***conjugate gradient method*** is a numerical and iterative method that, as we said, is used for symmetric positive definite matrices, especially for great sparse ones. It is assured to converge to the solution after a finite number of iterations if no roundup error is applied, but for this reason it is considered as a direct method. However, in reality, not rounding is almost impossible and inefficient.

Getting the formulation for this iterative method can be done in several ways. But in our case, we will use the optimization of convex functions. First, we need to look at quadrics, which we will get deeper into later in this chapter. For now, it is enough to say that a quadric is a polynomial of degree 2, and may be univariate or multivariate. Therefore, a quadric, which we will call $\boldsymbol{P}$, is usually presented like this:

$$P = \sum_{j=1}^{n}\sum_{i=1}^{n} a_{ij} \cdot x_i \cdot x_j + \sum_{k=1}^{n} b_k \cdot x_k + c$$

Where $\boldsymbol{x_1}, \boldsymbol{x_2}, \ldots, \boldsymbol{x_n}$ are the n different variables of the polynomials, and $\boldsymbol{a_{ij}}, \boldsymbol{b_k}$, and $\boldsymbol{c}$ are different constants that will depend on the quadric. If we look for help in linear algebra we would see that we can represent the quadric in the following form:

$$A = \begin{pmatrix} a_{11} & a_{12} & \cdots & a_{1n} \\ a_{21} & a_{22} & \cdots & a_{2n} \\ \vdots & \vdots & \cdots & \vdots \\ a_{n1} & a_{n2} & \cdots & a_{nn} \end{pmatrix}; \vec{x} = \begin{pmatrix} x_1 \\ x_2 \\ \vdots \\ x_n \end{pmatrix} \rightarrow P = \left(\overrightarrow{x^T} \cdot A \cdot \vec{x}\right) + \left(\overrightarrow{b^T} \cdot \vec{x}\right) + c$$

That even though we expressed it under matrix notation, a quadric is still represented by a scalar expression. That is how the above expression can be rewritten to express a quadric, but for our approach into the Conjugate Gradient method we will rewrite it just a little different. We will also name $\boldsymbol{P}$ a function of vector $\boldsymbol{x}$. Therefore,

$$P = f(\vec{x}) = \frac{1}{2} \cdot \overrightarrow{x^T} \cdot A \cdot \vec{x} - \left(\overrightarrow{b^T} \cdot \vec{x}\right) + c$$

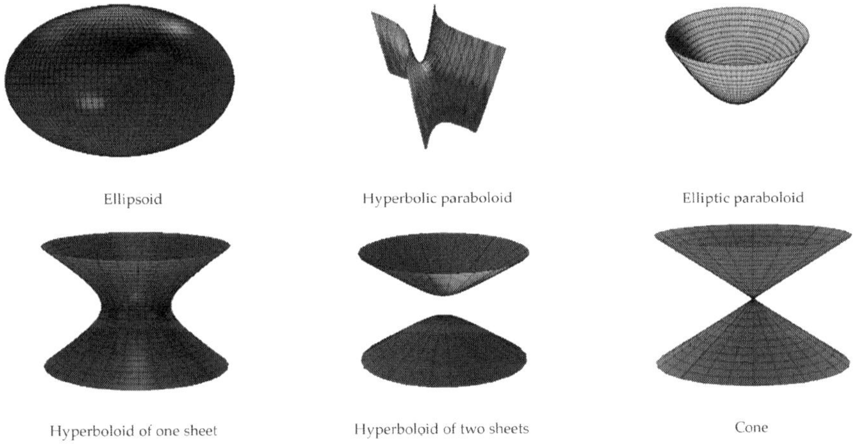

Figure 20: Examples of quadrics.

Source: hardycalculus.com

Figure 20 presents some examples of quadrics and their shapes. Going back to our topic, , we rewrote the expression for a reason. ***f*** is a function of vector ***x***, but it is a scalar function, which means that if we input a certain vector ***x***, function ***f*** will return a scalar. Suppose we want to find the minimum of ***f***. From calculus we know that in order to find the minimum we need to derive ***f*** with respect to each of its variables and equate those derivatives to zero which will give ***n*** equations that can be packed up using matrix notation. However, someone who has more experience in these area will know that this process is equivalent to calculating the gradient of ***f*** and setting it equal to zero. Independent of any method, we will get this:

$$\nabla f = \vec{0} \rightarrow \frac{1}{2}\left(A + A^T\right)\vec{x} - \vec{b} = \vec{0}$$

In case matrix ***A*** is symmetric, our gradient would then be:

$$\nabla f\left(\vec{x}\right) = A\vec{x} - \vec{b} = \vec{0}$$

We can see that minimizing a certain quadric ***f*** is equivalent to solving the problem $A\vec{x} = \vec{b}$. We just saw that matrix ***A*** must be symmetrical in order for the above to be accomplished, however, there is another condition. When deriving and setting that derivative to zero, we are able to get all the minimums in function ***f***, but the problem $A\vec{x} = \vec{b}$ has only one solution. Therefore, we only need a single minimum in ***f***. This condition requires that ***A*** be a positive definite matrix, because if ***A*** is positive definite, then ***f*** will have a bowl shape (the mathematical term is a convex function) and it will only have a single minimum. This can be seen in the following images: a paraboloid (for a ***definite function***), a parabolic cylinder (for a ***semi-definite function***), and a hyperbolic paraboloid (for an ***indefinite function***). Remember that all of these functions are equal to ***z***. So if we want to express them under matrix notation we need to do the following. Assuming vector ***x*** has the following form. Then,

$$\vec{x} = \begin{pmatrix} x \\ y \end{pmatrix}$$

$$z = x^2 + y^2 \rightarrow z = \overrightarrow{x^T}\begin{pmatrix} 1 & 0 \\ 0 & 1 \end{pmatrix}\vec{x} z = x^2 \rightarrow z = \overrightarrow{x^T}\begin{pmatrix} 1 & 0 \\ 0 & 0 \end{pmatrix}\vec{x} z = x^2 + y^2 \rightarrow z = \overrightarrow{x^T}\begin{pmatrix} 1 & 0 \\ 0 & -1 \end{pmatrix}\vec{x}$$

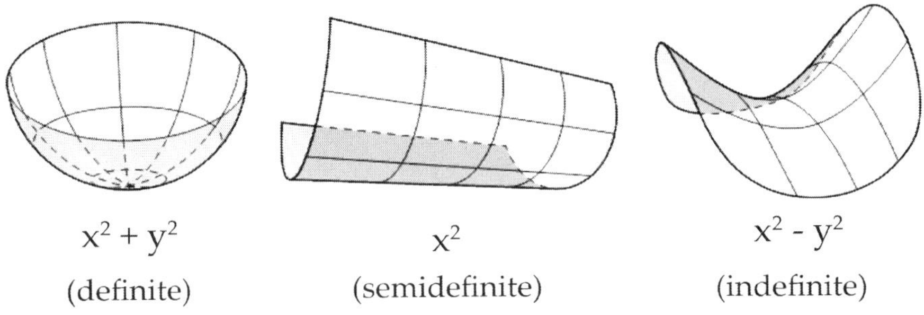

Figure 21: Definite, semi-definite, and indefinie quadrics.

We see that a definite function has a single minimum, a semi-definite function has a whole line of minimums, and an indefinite function has no minimum. If you want to know more about this, you will be going to another field of study called "convex optimization" which is not covered in this book.

Going back to our topic, we are looking for a solution to the problem $A\vec{x} = \vec{b}$ by minimizing a certain quadric, where ***A*** is a symmetric positive definite matrix. You must remember that vector ***b*** and the independent term ***c*** in our function ***f*** only affect where the minimum is, but they do not affect the shape of the function.

So instead of solving the problem $A\vec{x} = \vec{b}$ directly, we can stay with the problem of minimizing ***f*** and solving our original problem. Doing this may seem to transform our original problem into a more complex problem, but the methods like ***Steepest Descent*** and the ***Conjugate Gradient*** method are very easy to use and very intuitive. For this, let us start with the method of Steepest Descent.

Steepest Descent

This method is an iterative method that starts with an arbitrary point $\overrightarrow{x_0}$ and slides down our bowl-shaped function until it reaches another point near the minimum. This method is based on this principle because ***f*** has only 1 minimum. To do this, the method takes a series of steps in the direction that ***f*** decreases most quickly, which is the direction opposite to the gradient of ***f*** at that point $\overrightarrow{x_0}$. Therefore, the direction we are going to take is given by:

$$-\nabla f\left(\overrightarrow{x_0}\right) = b - A\overrightarrow{x_0}$$

It would be nice to introduce some definitions at this point. One of them is the ***error*** in our solution which corresponds to the difference between the approximate solution we have and the real solution which we do not have, that is, $\overrightarrow{e_i} = \overrightarrow{x_i} - \vec{x}$. In reality we will never have the true error term, because if we would have the error then we would also have the solution to our system; and if we have our solution, why would we be doing this?

In reality what we will have is another term called the ***residual*** which corresponds to the difference $\overrightarrow{r_i} = b - A\overrightarrow{x_i}$, which we can easily. Both terms, the ***error*** and the ***residual***, give us an idea of ***how far from the and which direction to take to get to the real solution***. Both are correlated like this: $\overrightarrow{r_i} = -A\overrightarrow{e_i}$. Therefore, from here we can also see $-\nabla f\left(\overrightarrow{x_i}\right) = \overrightarrow{r_i}$, where we can see that the residual is the direction in which ***f*** decreases most quickly.

By the way, from now on we will not write the arrow above vectors in our equations until the end of this subject. Now that we have this, we can write our iterative method and find the next point in our iterations like this:

$$x_{i+1} = x_i + \acute{a} \cdot r_i$$

Where ***alpha*** is a constant that corresponds to the length of the step we are taking. The question is how long should our step be? Well, first let us look at the following image which corresponds to the iso-lines or contour lines (colored in blue) of a convex quadric. We start at a point x_i (colored as a black dot) and move in the direction $-\nabla f$ (colored as a red vector) which is equal to the direction of the residual. Taking this direction with a certain step ***alpha*** (colored in black) will get us to a new point x_{i+1} (colored as a green dot). It is important to show that the negative of the gradient of ***f*** will not necessarily point at the direction of the minimum of ***f*** (this will be discussed further ahead), but with the conditions we pointed before (the positive definite condition), then it is assured that the negative of the gradient will point somewhere near the minimum.

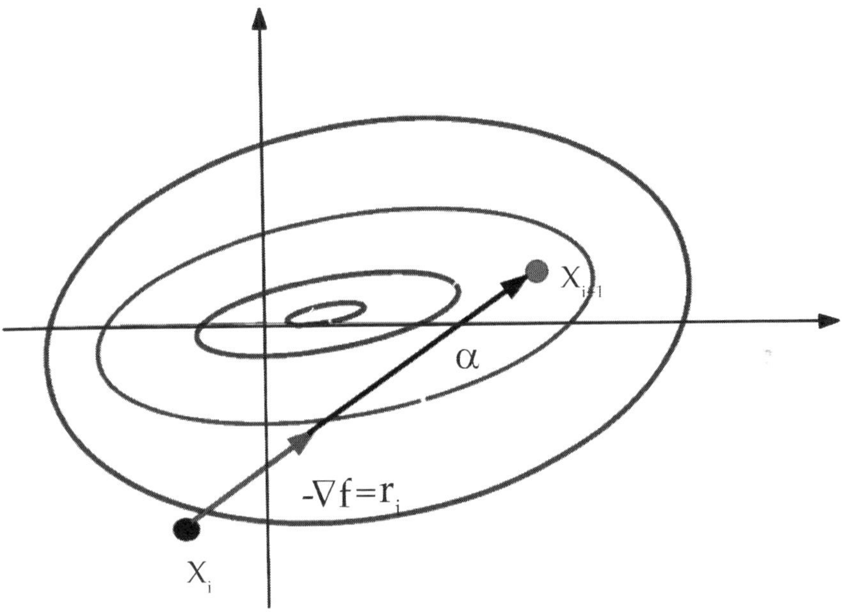

Figure 22: Iso-lines with Steepest Descent method.

Alpha needs to be a quantity that takes us the closest as possible to the minimum of ***f***. Therefore, we need to find ***alpha*** such that it minimizes ***f*** along the line corresponding to the direction of the residual, or in other words, we need to set the directional derivative $\frac{\partial}{\partial\alpha} f(x_{i+1})$ equal to zero. In calculus terms,

$$\frac{\partial}{\partial\alpha} f(x_{i+1}) = 0$$

By the chain rule we can express this as the following:

$$\frac{\partial}{\partial\alpha} f(x_{i+1}) = \left[\nabla f(x_{i+1})\right]^T \cdot \frac{d(x_{i+1})}{d\alpha}$$

The above derivative can be expressed as the product of two vectors, the transpose of the gradient of ***f*** at point x_{i+1} times the derivative of x_{i+1} with respect to alpha (which is also a vector). In order to calculate the last unknown derivative we remember that $x_{i+1} = x_i + \acute{a} \cdot r_i$, therefore its derivative will be $\frac{d(x_{i+1})}{d\alpha} = r_i$.. Remember that the residual is a vector. Therefore,

$$\left[\nabla f(x_{i+1})\right]^T \cdot r_i = 0$$

But we also know that the gradient of ***f*** at point x_{i+1} is equal to the negative of the residual r_{i+1}. From here we can see that

$$(-r_{i+1})^T \cdot r_i = 0 \rightarrow (r_{i+1})^T \cdot r_i = 0$$

which is a very important fact because every consecutive pair of residuals will be orthogonal to each other. But we still have to look for an expression for ***alpha***. Note that:

$$r_{i+1} = b - A \cdot x_{i+1} = b - A \cdot (x_i + \alpha \cdot r_i) = r_i - \alpha \cdot A \cdot r_i$$

If we input this in our last equation and solve for alpha we get:

$$(r_i - \alpha \cdot A \cdot r_i)^T \cdot r_i = 0$$

$$\alpha = \frac{r_i^T \cdot r_i}{(A \cdot r_i)^T \cdot r_i}$$

And here we have it, the method of Steepest Descent. Putting it all together we get

$$r_i = b - A \cdot x_i$$

$$\alpha_i = \frac{r_i^T \cdot r_i}{(A \cdot r_i)^T \cdot r_i}$$

$$x_{i+1} = x_i + \alpha_i \cdot r_i$$

A common behavior with the residuals of Steepest Descent is the zig-zag behavior because each pair of consecutive residuals are orthogonal to each other:

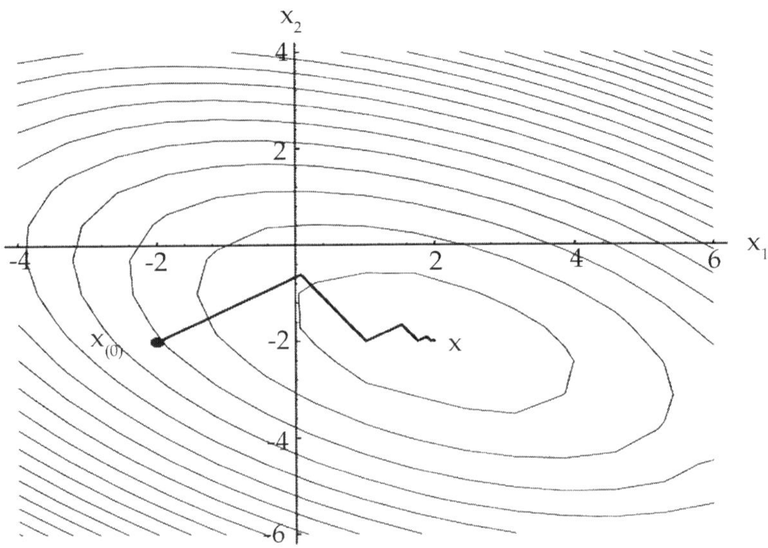

Figure 23: Steepest Descent method behavior.

Source: Painless Conjugate Gradient method

The algorithm written above requires two matrix vector multiplications per iteration. The computational cost of Steepest Descent is dominated by matrix-vector products. However, we can minimize this cost by calculating each residual with respect to the previous one like this. Taking the last equation in the algorithm written above:

$$x_{i+1} = x_i + á_i \cdot r_i$$

Subtracting the real solution x from both sides

$$x_{i+1} - x = x_i - x + á_i \cdot r_i$$

$$e_{i+1} = e_i + á_i \cdot r_i$$

Multiplying both sides by A

$$-r_{i+1} = -r_i + á_i \cdot A \cdot r_i$$

Rearranging

$$r_{i+1} = r_i - á_i \cdot A \cdot r_i$$

The algorithm can be simplified like this:

$r_0 = b - A \cdot x_0$

$$\alpha_i = \frac{r_i^T \cdot r_i}{(A \cdot r_i)^T \cdot r_i}$$

$r_{i+1} = r_i - \alpha_i \cdot A \cdot r_i$

Even though doing both approaches is the same, the disadvantage of using this last recurrence is that, as the sequence of residuals are generated without taking the actual residual $r_i = b - Ax_i$, there might be an accumulation of roundup error giving a point near $\boldsymbol{x}$ as the final solution. This problem can be solved by taking the real residual $r_i = b - Ax_i$ after a finite number of iterations to eliminate the roundup error. And after the residual becomes very small we can calculate the solution like this:

$x \approx x_0 + \sum á_i \cdot r_i$

Now that we have talked about the method, we also need to talk about its convergence because we want to know how fast our method is or how good it is. For this, we will discuss something that we have mentioned before -- that the residual usually does not have the same direction as the error. It always happens unless the error is an eigenvector of matrix A. Remember that the residual and the error are correlated like this:

$r_i = -A \cdot e_i$

And both vectors will have the same direction if the error is an eigenvector of $\boldsymbol{A}$. In this case, convergence is immediate, since with only one iteration we have convergence to the exact solution. This is demonstrated by assuming that the error is an eigenvector of $\boldsymbol{A}$ (remember that if the error is an eigenvector then the residual is also an eigenvector). Assuming this we get:

$if A \cdot e_i = \lambda_e \cdot e_i = -r_i \rightarrow$

$$e_{i+1} = e_i + \alpha_i \cdot r_i$$

$$e_{i+1} = e_i + \frac{r_i^T \cdot r_i}{(A \cdot r_i)^T \cdot r_i} \cdot r_i$$

$$e_{i+1} = e_i + \frac{r_i^T \cdot r_i}{\lambda_e \cdot r_i^T \cdot r_i} \cdot (-\lambda_e \cdot e_i)$$

$$e_{i+1} = 0$$

But this case only happens if the error is really an eigenvector. In order to get a more detailed analysis we must express the error term as a linear combination of eigenvectors. Remember that because ***A*** is a symmetric positive definite matrix, then its eigenvalues are positive and its eigenvectors are orthogonal. Therefore, if we have a certain field of scalars c_1, c_2, …, c_n, name the different eigenvalues of ***A*** λ_1, λ_2, …, λ_n, the different eigenvectors of ***A*** v_1, v_2, …, v_n, and assuming they are orthonormal, then we can express the error term as:

$$e_i = \sum_{j=1}^{n} c_j \cdot v_j$$

And from here we can get the following expressions:

$$\| e_i \|^2 = e_i^T \cdot e_i = \sum_{j=1}^{n} c_j^2$$

$$(A \cdot e_i)^T \cdot e_i = \left(\sum_{j=1}^{n} c_j \cdot v_j^T \right) \cdot \left(\sum_{j=1}^{n} c_j \cdot \lambda_j \cdot v_j^T \right) = \sum_{j=1}^{n} c_j^2 \cdot \lambda_j$$

$$\| r_i^2 \| = r_i^T \cdot r_i = \sum_{j=1}^{n} c_j^2 \cdot \lambda_j^2$$

$$(A \cdot r_i)^T \cdot r_i = \sum_{j=1}^{n} c_j^2 \lambda_j^3$$

To prove convergence we need to prove that the error term gets smaller after each iteration. This is done by the norm of the error vector.

However, working with the Euclidean norm is not so comfortable. It is better to work with the energy norm which is defined like this:

$$\| e_i \|_A = \left[\left(A \cdot e_i \right)^T \cdot e_i \right]^{1/2}$$

Therefore, with this norm we can get the following expression which relates the future norm of the error vector with respect to the present norm (remember that **A** is a symmetric matrix while deriving the expression):

$$\| e_{i+1} \|_A^2 = \left(A \cdot e_{i+1} \right)^T \cdot e_{i+1}$$

$$\| e_{i+1} \|_A^2 = \left(A \cdot (e_i + \alpha_i r_i) \right)^T \cdot (e_i + \alpha_i r_i)$$

$$\| e_{i+1} \|_A^2 = \left(A \cdot e_i \right)^T \cdot e_i + 2 \cdot \alpha_i \cdot r_i^T \cdot \left(A \cdot e_i \right) + \alpha_i^2 \cdot \left(A \cdot r_i \right)^T \cdot r_i$$

$$\| e_{i+1} \|_A^2 = \| e_i \|_A^2 + 2 \cdot \frac{r_i^T \cdot r_i}{\left(A \cdot r_i \right)^T \cdot r_i} \cdot r_i^T \cdot \left(-r_i \right) + \left(\frac{r_i^T \cdot r_i}{\left(A \cdot r_i \right)^T \cdot r_i} \right)^2 \cdot \left[\left(A \cdot r_i \right)^T \cdot r_i \right]$$

$$\| e_{i+1} \|_A^2 = \| e_i \|_A^2 - 2 \cdot \frac{\left(r_i^T \cdot r_i \right)^2}{\left(A \cdot r_i \right)^T \cdot r_i} + \frac{\left(r_i^T \cdot r_i \right)^2}{\left(A \cdot r_i \right)^T \cdot r_i}$$

$$\| e_{i+1} \|_A^2 = \| e_i \|_A^2 - \frac{\left(r_i^T \cdot r_i \right)^2}{\left(A \cdot r_i \right)^T \cdot r_i}$$

$$\| e_{i+1} \|_A^2 = \| e_i \|_A^2 \cdot \left(1 - \frac{\left(r_i^T \cdot r_i \right)^2}{\left[\left(A \cdot r_i \right)^T \cdot r_i \right] \left[\left(A \cdot e_i \right)^T \cdot e_i \right]} \right)$$

Finally, we got an expression for the norm of the future error in terms of the present error. From the above equation we can see that the algorithm will converge if:

$$1-\frac{\left(r_i^T \cdot r_i\right)^2}{\left[\left(A \cdot r_i\right)^T \cdot r_i\right]\left[\left(A \cdot e_i\right)^T \cdot e_i\right]}<|1|$$

If we get the above equation in terms of the norms we calculated before and we call it ***omega***, we get:

$$\omega=1-\frac{\left(\sum_{j=1}^{n} c_j^2 \cdot \lambda_j^2\right)^2}{\left(\sum_{j=1}^{n} c_j^2 \cdot \lambda_j^3\right) \cdot\left(\sum_{j=1}^{n} c_j^2 \cdot \lambda_j\right)}$$

The above expression is what gives the rate of convergence of the Steepest Descent method, since as ***omega*** becomes ***smaller*** the method will converge ***faster***. Omega will always be between 0 and 1, proving why it is a bit trickier. However, taking the case when we have two eigenvectors gives an idea on how the rate of convergence is affected by the different terms. In case we have two eigenvectors, then

$$\omega=1-\frac{\left(c_1^2 \lambda_1^2+c_2^2 \lambda_2^2\right)^2}{\left(c_1^2 \lambda_1^3+c_2^2 \lambda_2^3\right) \cdot\left(c_1^2 \lambda_1+c_2^2 \lambda_2\right)}$$

Assuming that $\lambda_1 \geq \lambda_2$ and introducing the factors $k=\lambda_1 / \lambda_2$ and $\mu=c_2 / c_1$, then we have a friendlier expression:

$$\omega=1-\frac{\left(k^2+\mu^2\right)^2}{\left(k^3+\mu^2\right) \cdot\left(k+\mu^2\right)}$$

In this case, when any or both terms, ***k*** or μ, are small then convergence is almost immediate. However, when both of them are large, there is convergence but slower. The worst case is when $\mu=\pm k$. Therefore, if we introduce this worst condition in the above formula, then we have

$$\omega=\left(\frac{k-1}{k+1}\right)^2 \rightarrow \textit{for worst condition}$$

wherein omega is always less than 1, but gets closer and closer to one as ***k*** grows to infinity. Take note that ***k*** is usually defined as the ratio between the greatest and the smallest eigenvalue and it represents the condition number of matrix $\boldsymbol{A}$. For those who do not know, the ***condition number*** is a number that indicates how ill-conditioned a matrix is, that is, how closer it is to being not invertible, which translates to a system having no solution. The higher the condition number is, the harder it is to solve the problem $A\vec{x} = \vec{b}$, and the harder it is for the method to converge.

However, we have shown that the Steepest Descent method converges for every starting point and for every positive definite symmetric matrix. This method is a very good approximation to usual problems, but it can get better, and with today's problems and matrix of thousands of rows and columns it is good to optimize algorithms even more. To optimize this algorithm we must now go to the ***Conjugate Directions*** method.

Conjugate Directions

The conjugate gradient method comes from the desire of optimizing the Steepest Descent method. The idea mathematicians had when working with the Steepest Descent method was that, even when consecutive residuals were orthogonal to each other, many residuals had the same direction than another one that came before. Therefore, they looked for a way such that all directions with the residuals were "orthogonal" to each other, not only to the previous one, that is, there is only a finite number of directions. In other words, after a finite number of iterations the problem is solved. For this reason, the method of Conjugate Gradient is sometimes called a direct method, because it gives the real answer after a finite number of directions, however, in reality it is an iterative method.

But in order to get all the "orthogonal" directions, we need to study the Conjugate Directions method first, in which relies this principle. For this suppose we have a certain set of orthogonal directions d_i and that our iterative method is established like this:

$$x_{i+1} = x_i + \alpha_i \cdot d_i$$

If we take advantage the all directions must be orthogonal, then also all future errors must be orthogonal. With this idea we can get an expression for ***alpha***:

$$d_i^T \cdot e_{i+1} = 0$$

$$d_i^T \cdot (e_i + \alpha_i \cdot d_i) = 0$$

$$\alpha_i = -\frac{d_i^T \cdot e_i}{d_i^T \cdot d_i}$$

Unfortunately this means nothing, because we cannot compute ***alpha*** without knowing the error ***e***; and if we know the error, the problem would have already been solved. To proceed with this problem we do not make the directions orthogonal, but we make them ***A orthogonal***. This concept means the following: Two directions are orthogonal if $d_i^T \cdot d_j = 0$ for any ***i*** different than ***j***, however, two directions are ***A*** orthogonal if $d_i^T \cdot A \cdot d_j = 0$ for every ***i*** different than ***j***. **A orthogonality**, geometrically means the following:

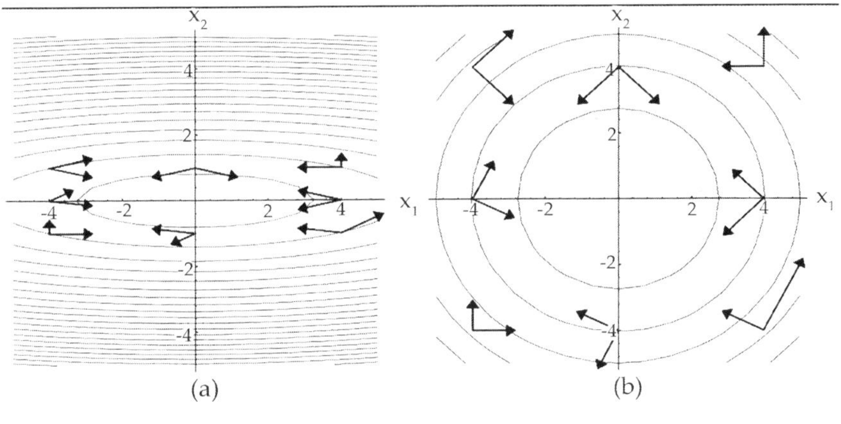

Figure 24: A-orthogonal vector.

Source: Painless Conjugate Gradient method

On the left side of the image we have a certain quadric represented by a matrix ***A***, with a certain set of vectors. However, if we "stretch" the quadric until it becomes circular, like the one in the right side of the image, we see that those set of vectors are now orthogonal. This is the meaning of being ***A*** orthogonal vectors. It is not a coincidence that this condition is equivalent to finding the minimum point of f along the direction d_i, like in Steepest Descent. To see this, we make the following:

$$\frac{\partial}{\partial\alpha} f(x_{i+1}) = \left[\nabla f(x_{i+1})\right]^T \cdot \frac{d(x_{i+1})}{d\alpha}$$

$$(-r_{i+1})^T \cdot d_i = 0 \rightarrow (r_{i+1})^T \cdot d_i = 0$$

The last expressions means that the future residual is orthogonal to the present direction, but getting deeper into our case we get:

$$r_{i+1} = b - A\cdot x_{i+1} = b - A\cdot(x_i + \alpha\cdot d_i) = r_i - \alpha\cdot A\cdot d_i$$

And using the same derivation than with Steepest Descent we find that:

$$(r_i - \alpha\cdot A\cdot d_i)^T \cdot d_i = 0$$

$$\alpha_i = \frac{r_i^T \cdot d_i}{(A\cdot d_i)^T \cdot d_i}$$

In case the directions we are talking about would be the residuals, then this would be equivalent to using the Steepest Descent method. It is important to notice that we still have not derived an algorithm to obtain the ***A*** orthogonal directions. For now let us stay with the idea that there exists a set of ***A*** orthogonal directions that we still have not found yet.

Even though we still have not found them, this idea is very important, because it is the idea that allows us to calculate the error term in a finite number of steps. To prove this, let us express the initial error term as a linear combination of the ***A*** orthogonal directions d_j:

$$e_0 = \sum_{j=0}^{n-1} c_j d_j$$

We need to find the values of the different constants c_j in order to express our error correctly. We do this with the idea that the directions are ***A*** orthogonal, therefore, if we multiply the above expression by $d_k^T \cdot A$ we would get

$$d_k^T \cdot A \cdot e_0 = \sum_{j=0}^{n-1} c_j \cdot d_k^T \cdot A \cdot d_j$$

$$d_k^T \cdot A \cdot e_0 = c_k \cdot d_k^T \cdot A \cdot d_k$$

$$c_k = \frac{d_k^T \cdot A \cdot e_0}{d_k^T \cdot A \cdot d_k}$$

Where this can also be changed a bit to get a very useful relation. Let us use the fact that the different directions are ***A*** orthogonal again by rearranging the above expression a little bit:

$$c_k = \frac{d_k^T \cdot A \cdot \left(e_0 + \sum_{i=0}^{k-1} \alpha_i \cdot d_i\right)}{d_k^T \cdot A \cdot d_k} - \frac{\alpha_i \cdot d_k^T \cdot A \cdot \sum_{i=0}^{k-1} d_i}{d_k^T \cdot A \cdot d_k}$$

The second term in the right side of the above equation is zero. Therefore, we would get the following:

$$c_k = \frac{d_k^T \cdot A \cdot e_k}{d_k^T \cdot A \cdot d_k} = -\frac{d_k^T \cdot r_k}{d_k^T \cdot A \cdot d_k} = -á_k$$

Which gives a new way of looking at the error, since we see that the different values of ***alpha***, the steps we have to make to minimize our function ***f*** along each direction, are equal to the different components of our initial error vector. This means that after each iteration we are cutting the error term component by component, meaning that after ***n*** iterations, every component is cut away and $e_n = 0$ meaning that this method actually does work in a finite number of iterations. The above property also let us express the error term as the following way:

$$e_i = \sum_{j=i}^{n-1} c_j \cdot d_j$$

The only thing we need for this method to start working are the directions. For this we use the ***Conjugate Gram Schmidt Process***. We already saw the regular Gram Schmidt process, however, this one is just a little different. As we know, for Gram Schmidt do its job, we need a set of ***n*** linearly independent vectors, which we will name $\boldsymbol{u_0}$, $\boldsymbol{u_2}$, …, $\boldsymbol{u_{n-1}}$, which may be the directions of the different coordinate axes. Then we need to take one vector and subtract from it any components that are not **A** orthogonal to the previous vectors.

$$d_i = u_i - \sum_{k=0}^{i-1} c_{ik} \cdot d_k$$

where the different terms $c_{ik} \cdot d_k$ are the different components we need to cut from vector u_i to get a new **A** orthogonal direction. In order to find the different values of c_{ik} we multiply the above expression by $d_j^T \cdot A$ at both sides with $j \neq i$ obtaining the following:

$$d_j^T \cdot A \cdot d_i = d_j^T \cdot A \cdot u_i - \sum_{k=0}^{i-1} c_{ik} d_j^T \cdot A \cdot d_k$$

$$0 = d_j^T \cdot A \cdot u_i - c_{ij} \cdot d_j^T \cdot A \cdot d_j$$

$$c_{ij} = \frac{d_j^T \cdot A \cdot u_i}{d_j^T \cdot A \cdot d_j}$$

This is the way that we may look for all the conjugate directions. Therefore, we can now express the method of Conjugate Directions: Having a set of ***n*** linearly independent vector u_i:

$for 1 \leq i \leq n$

$$\alpha_i = \frac{r_i^T \cdot d_i}{\left(A \cdot d_i\right)^T \cdot d_i}$$

$$d_i = u_i - \sum_{k=0}^{i-1} c_{ik} \cdot d_k$$

$$c_{ij} = \frac{d_j^T \cdot A \cdot u_i}{d_j^T \cdot A \cdot d_j} \; for j = 1, 2, \ldots, i-1$$

$$r_i = b - A \cdot x_i$$

$continuation \rightarrow$

$$x_{i+1} = x_i + \alpha_i \cdot d_i$$

Remember that c_{ij} can be considered as the elements in a certain matrix. You may have noticed that calculating these elements takes a lot of effort and calculation time. The Conjugate Directions method can also be improved by changing the last equation like this: $r_{i+1} = r_i - á_i \cdot A \cdot d_i$ and calculating the residual as a linear combination of other residuals and the last known direction.

Now everything is in place for the Conjugate Gradient method, because the Conjugate Gradient method is the union between the Steepest Descent and the Conjugate Directions method. This means that the u_i vectors we are going to use to make $\boldsymbol{A}$ orthogonal and build our directions, are going to be the residuals. For this we need to look at some properties of Conjugate Directions. The first property is that all present residuals are orthogonal to all past directions. To show this, we need to multiply the error term at the $\boldsymbol{i^{th}}$ step by $d_k^T \cdot A$ with $k < i$. Therefore,

$$e_i = \sum_{j=i}^{n-1} c_j \cdot d_j$$

$$d_k^T \cdot A \cdot e_i = \sum_{j=i}^{n-1} c_j \cdot d_k \cdot A \cdot d_j$$

$$d_k^T \cdot r_i = 0 \rightarrow k < i$$

This is an important advancement because this will help us show that each residual will be orthogonal to all the ones that came before, not only to the last one like in Steepest Descent, which means that after $\boldsymbol{n}$ iterations we will be out of orthogonal directions, that we already have reached the solution. This last statement can be proved by multiplying the Conjugate Gram Schmidt Process by r_j^T with $j > i$, and concluding that

$$r_j^T \cdot d_i = r_j^T \cdot u_i - \sum_{k=0}^{i-1} c_{ik} \cdot r_j^T \cdot d_k$$

$$r_j^T \cdot u_i = 0$$

And as in the Conjugate Gradient method, the vectors we are using are the residuals then all the residuals will be orthogonal to each other:

$$r_j^T \cdot r_i = 0 \rightarrow j \neq i$$

We can also demonstrate another useful relationship if we multiply the Conjugate Gram Schmidt process by r_i^T to get:

$$r_i^T \cdot d_i = r_i^T \cdot u_i - \sum_{k=0}^{i-1} c_{ik} \cdot r_i^T \cdot d_k$$

$$r_i^T \cdot d_i = r_i^T \cdot u_i = r_i^T \cdot r_i$$

You might probably be wondering why we did all these. The answer is that, by taking the residuals as vectors for constructing the ***A*** orthogonal directions, then all is simplified, a lot, and this relationships will help us know why. For now, the most important improvement is that we found that if we take the residuals as u_i vectors then all the residuals are orthogonal. However, we can simplify the method even more by looking at elements c_{ij}. One equation we derived before was this:

$$r_{i+1} = r_i - \alpha_i \cdot A \cdot d_i$$

If we multiply both sides by another residual r_j^T, and introduce alpha in the equation we get:

$$r_j^T \cdot r_{i+1} = r_j^T \cdot r_i - \frac{r_i^T \cdot d_i}{(A \cdot d_i)^T \cdot d_i} \cdot r_j^T \cdot A \cdot d_i$$

We can relate this equation with the expression for matrix c_{ji} (the order of ***i*** and ***j*** are flipped, but that does not matter as we shall soon see), like this:

$$r_j^T \cdot r_{i+1} = r_j^T \cdot r_i - \left(\frac{r_j^T \cdot A \cdot d_i}{(A \cdot d_i)^T \cdot d_i} \right) \cdot r_i^T \cdot d_i$$

$$r_j^T \cdot r_{i+1} = r_j^T \cdot r_i - c_{ji} \cdot r_i^T \cdot d_i$$

$$c_{ji} = \frac{r_j^T \cdot r_i - r_j^T \cdot r_{i+1}}{r_i^T \cdot d_i} = \frac{r_j^T \cdot r_i - r_j^T \cdot r_{i+1}}{r_i^T \cdot r_i}$$

Where we see that most of the elements of the above expression will be zero depending on the values of ***i*** and ***j***. Let us look at this in the following way:

$$for j = i \rightarrow c_{ii} = \frac{r_i^T \cdot r_i}{r_i^T \cdot r_i} = 1$$

$$for j = i+1 \rightarrow c_{i,i-1} = -\frac{r_i^T \cdot r_i}{r_{i-1}^T \cdot r_{i-1}}$$

$$otherwise \rightarrow c_{ij} = 0$$

Let us take a deep breath and talk about what just happened. We had the Conjugate Directions method, which is an improvement of the Steepest Descent method, by taking a set of vectors $u_1, u_2, ..., u_n$ and transforming them into a set of ***n A*** orthogonal directions. With the Conjugate Directions method we knew that convergence to the exact solution will happen after ***n*** steps, and to make the ***A*** orthogonal directions we used the Conjugate Gram Schmidt Process which involved making many calculations because of the elements c_{ij}.

The Conjugate Gradient method takes the u_i vectors as the residuals. By doing this we saw that all residuals calculated in the algorithm will be orthogonal and that the calculation of elements c_{ij} are simplified a lot. If we input this conditions we get the Conjugate Gradient method, which we will now present, step by step:

- First we need to start with a residual, which comes from our seed vector x_0. This residual will also be equal to the first direction we are taking:

$$r_0 = b - A \cdot x_0 = d_0$$

- With this we find the next residuals like this:

$$\alpha_i = \frac{r_i^T \cdot r_i}{\left(A \cdot d_i\right)^T \cdot d_i} \quad x_{i+1} = x_i + \alpha_i \cdot d_i \quad r_{i+1} = b - A \cdot x_i$$

- With this residuals we calculate the next direction we are going to take. In this case look at how the algorithm for finding this direction is simplified from the conditions we found for the elements c_{ij}.

$$d_i = u_i - \sum_{k=0}^{i-1} c_{ik} \cdot d_k = r_i - c_{i,i-1} \cdot d_{i-1}$$

Note that as we will only have one element from matrix c_{ij} in each of our equations then we can calculate the directions like this:

$$c_{i+1} = \frac{r_{i+1}^T \cdot r_{i+1}}{r_i^T \cdot r_i} \quad d_{i+1} = r_{i+1} + c_{i+1} \cdot d_i$$

And then we would repeat the algorithm until we get the correct answer. A little example of this method is shown next. Suppose we have the following system of equations, which has a positive definite symmetric matrix.

$$\begin{pmatrix} 5 & 1 & 2 \\ 1 & 4 & 3 \\ 2 & 3 & 7 \end{pmatrix} \begin{pmatrix} x \\ y \\ z \end{pmatrix} = \begin{pmatrix} 3 \\ 4 \\ 5 \end{pmatrix}$$

If we apply the Conjugate Gradient method starting with our seed vector as the zero vector we would get:

$$step0$$

$$r_0 = d_0 = b - A \cdot x_0 = \begin{pmatrix} 3 \\ 4 \\ 5 \end{pmatrix}$$

$$\alpha_0 = 0.1024$$

$$step1 \qquad\qquad step2$$

$$x_1 = x_0 + a_0 \cdot d_0 = \begin{pmatrix} 0.3073 \\ 0.4098 \\ 0.5122 \end{pmatrix} \qquad x_2 = \begin{pmatrix} 0.3333 \\ 0.6666 \\ 0.3333 \end{pmatrix}$$

$$r_1 = \begin{pmatrix} 0.0286 \\ 0.5163 \\ -0.4303 \end{pmatrix} \qquad r_2 = \begin{pmatrix} 0 \\ 0 \\ 0 \end{pmatrix}$$

$$c_1 = 0.0090 \qquad c_2 = 0$$

$$d_1 = \begin{pmatrix} 0.0558 \\ 0.5526 \\ -0.3850 \end{pmatrix} \qquad d_2 = \begin{pmatrix} 0 \\ 0 \\ 0 \end{pmatrix}$$

$$\alpha_1 = 0.4647 \qquad \alpha_2 = NaN$$

Where we see convergence after 2 steps, if no roundup error is made during the calculations. Truly a magnificent algorithm. This method is remarkable and is famous for solving this kinds of problems. You can check and see for yourself that it does not matter with what vector seed we start as we will always get convergence after three steps or less. The process is very interesting to see if our seed vector is very far from the solution.

SINGULAR VALUE DECOMPOSITION

As this chapter is about important topics in Algebra, we will give also special attention to this topic, since it has been growing in fame for the last two decades. The ***singular value decomposition*** is probably the most famous and important decomposition a matrix can have. It is used nowadays in fields that go from statistics to medicine, from information storage to facial recognition software, from engineering to game development.

The singular value decomposition, or as many people call it, the ***SVD*** may sound like a marvelous approach, and in reality it is, but let us take a look why. In the previous chapter we saw that a symmetric matrix, let us call it ***A***, can be factored like this:

$$A = S\Lambda S^T$$

But this definition comes from the fact that ***S*** is an orthogonal matrix formed from the eigenvectors. However, why is this factorization only applicable to symmetric matrices? Why can't this be made to common matrices? Well, indeed it can be done to matrices that are not symmetric. In reality, the singular value decomposition can be done to any matrix. So let us state our approach.

The ***Singular Value Decomposition*** is a decomposition made with the following idea. Suppose we have a matrix ***A***, and we multiply our matrix ***A*** by a given unit vector ***v***. This multiplication will give a certain vector ***u*** that will possibly not have the same direction as ***v***, and also may not be a unit vector. But let us suppose we normalize vector ***u*** and multiply it by a constant sigma so that we do not change the equality. Therefore,

$$A\hat{v} = \vec{u} \rightarrow A\hat{v} = \sigma\hat{u}$$

Now let us take our problem even further, suppose we want to find a whole set of ortho-normal vectors ***v***, such that they are able to give a whole set of orthogonal vectors ***u***. It is important to notice that vectors ***u*** are not imposed to be unit vectors but we are able to normalize them and add a constant to the system, thereby getting

$$A\left(\widehat{v_1}, \widehat{v_2}, \ldots, \widehat{v_n}\right) = \left(\sigma_1\widehat{u_1}, \sigma_2\widehat{u_2}, \ldots, \sigma_n\widehat{u_n}\right)$$

We are able to get this into matrix notation like the following:

$$AV = U \cdot \Sigma$$

Where ***U*** is an orthonormal matrix and ***Σ*** is a diagonal matrix. Since matrix ***V*** is formed by a set of orthonormal vectors then we can solve for ***A*** by multiplying the right side by the transpose of ***V***. Therefore, we would get:

$$A = U \cdot \Sigma \cdot V^T$$

We have just found a new decomposition to matrix ***A*** without implying any condition of symmetry, size, rank, or anything. This is the so called Singular Value Decomposition. However, the question arises. How do we compute it? The answer to this question is easy. For this we need to do the following arrangements. First, we need to transpose the above equation and get this:

$$A = U \cdot \Sigma \cdot V^T$$

$$A^T = V \cdot \Sigma \cdot U^T$$

Remember that matrix ***Σ*** is a diagonal matrix, therefore, it is symmetric. But what happens if we multiply ***A*** by its transpose? This happens:

$$A \cdot A^T = U \cdot \Sigma \cdot V^T \cdot V \cdot \Sigma \cdot U^T = U \cdot \Sigma \cdot \left(V^T \cdot V\right) \cdot \Sigma \cdot U^T = U \cdot \Sigma^2 \cdot U^T$$

Remember that ***V*** is an orthonormal matrix, therefore, its transpose is its inverse. We know that ***A*** times its transpose is a symmetric matrix. We also know that it will have other properties if ***A*** is square and full rank. However, let us focus on symmetry. Matrix ***A*** by its transpose is symmetric, therefore, it has all positive eigenvalues and all of its eigenvectors are orthogonal. Thus, it has an eigendecomposition. But let us take a closer look. We just saw that:

$$A \cdot A^T = U \cdot \Sigma^2 \cdot U^T$$

We can see that the above expression corresponds exactly to the eigendecomposition of $A \cdot A^T$ because it is an orthogonal matrix times a diagonal matrix times the transpose of the orthogonal matrix. From this conclusion we can see that matrix ***U*** is formed by the eigenvectors of $A \cdot A^T$, and matrix ***Σ*** is formed by the square roots of its eigenvalues. But we still need to calculate matrix ***V*** in order to complete the problem. For this we do the following multiplication:

$$A^T \cdot A = V \cdot \Sigma \cdot U^T \cdot U \cdot \Sigma \cdot V^T = V \cdot \Sigma \cdot \left(U^T \cdot U\right) \cdot \Sigma \cdot V^T = V \cdot \Sigma^2 V^T$$

Here, we see a similar approach. $A^T \cdot A$ is a symmetric matrix and therefore, also has an eigendecomposition. We see that matrix ***V*** is a matrix formed by the eigenvectors of $A^T \cdot A$, and again, ***Σ*** is a diagonal matrix formed by the square roots of the eigenvalues of A^T. T herefore, we now have a way to compute for all components necessary for the

Singular Value Decomposition. Let us compute some examples to see how this works.

Suppose we have the following matrix ***A***, and we want to compute its Singular Value Decomposition:

$$A = \begin{pmatrix} 1 & 2 & 3 \\ 5 & 9 & 7 \\ 4 & 7 & 4 \end{pmatrix}$$

From what we just saw in the theory, we need to compute the following:

$$A \cdot A^T = U \cdot \Sigma^2 \cdot U^T$$

$$A^T \cdot A = V \Sigma^2 \cdot V^T$$

Therefore,

$$A \cdot A^T = \begin{pmatrix} 14 & 44 & 30 \\ 44 & 155 & 111 \\ 30 & 111 & 81 \end{pmatrix}$$

If we calculate the eigenvalues and eigenvectors we will find that they are:

$$\lambda_1 = 0; \lambda_2 = 247.16; \lambda_3 = 2.84$$

$$\vec{e_1} = \begin{pmatrix} 1 \\ -1 \\ 1 \end{pmatrix}; \vec{e_2} = \begin{pmatrix} 0.3912 \\ 1.3912 \\ 1 \end{pmatrix}; \vec{e_3} = \begin{pmatrix} -1.3415 \\ -0.3415 \\ 1 \end{pmatrix}$$

But we need to normalize them. Thus, they become

$$\widehat{e_1} = \begin{pmatrix} 0.5774 \\ -0.5774 \\ 0.5774 \end{pmatrix}; \widehat{e_2} = \begin{pmatrix} 0.2226 \\ 0.7916 \\ 0.5690 \end{pmatrix}; \widehat{e_3} = \begin{pmatrix} -0.7856 \\ -0.2000 \\ 0.5856 \end{pmatrix}$$

Therefore, our matrix ***U*** and ***Σ*** will be

$$U = \begin{pmatrix} 0.5774 & 0.2226 & -0.7856 \\ -0.5774 & 0.7916 & -0.2000 \\ 0.5774 & 0.5690 & 0.5856 \end{pmatrix}; \Sigma = \begin{pmatrix} \sqrt{\lambda_1} & 0 & 0 \\ 0 & \sqrt{\lambda_2} & 0 \\ 0 & 0 & \sqrt{\lambda_3} \end{pmatrix}$$

Remember that each eigenvalue must go in order with its corresponding eigenvector. Now we need to calculate matrix ***V***.

$$A^{T} \cdot A = \begin{pmatrix} 42 & 75 & 54 \\ 75 & 134 & 97 \\ 54 & 97 & 74 \end{pmatrix}$$

wherein

$$\lambda_1 = 0; \lambda_2 = 247.16; \lambda_3 = 2.84$$

$$\vec{e_1} = \begin{pmatrix} 13 \\ -8 \\ 1 \end{pmatrix}; \vec{e_2} = \begin{pmatrix} 0.7609 \\ 1.3615 \\ 1 \end{pmatrix}; \vec{e_3} = \begin{pmatrix} -0.3935 \\ -0.5145 \\ 1 \end{pmatrix}$$

But we need to normalize them. Thus,

$$\vec{e_1} = \begin{pmatrix} 0.8498 \\ -0.5230 \\ 0.0654 \end{pmatrix}; \vec{e_2} = \begin{pmatrix} 0.4107 \\ 0.7349 \\ 0.5397 \end{pmatrix}; \vec{e_3} = \begin{pmatrix} -0.3303 \\ -0.4318 \\ 0.8393 \end{pmatrix}$$

Therefore, our matrix ***V*** and ***Σ*** will be

$$V = \begin{pmatrix} 0.8498 & 0.4107 & -0.3303 \\ -0.5230 & 0.7349 & -0.4318 \\ 0.0654 & 0.5397 & 0.8393 \end{pmatrix}; \Sigma = \begin{pmatrix} \sqrt{\lambda_1} & 0 & 0 \\ 0 & \sqrt{\lambda_2} & 0 \\ 0 & 0 & \sqrt{\lambda_3} \end{pmatrix}$$

Again, each eigenvector must be in order with its respective eigenvalue. Having calculated all the necessary components, we can now write the Singular Value Decomposition of our matrix ***A***. Therefore,

$$A = U \cdot \Sigma \cdot V^{T}$$

$$A = \begin{pmatrix} 0.5774 & 0.2226 & -0.7856 \\ -0.5774 & 0.7916 & -0.2000 \\ 0.5774 & 0.5690 & 0.5856 \end{pmatrix} \cdot \begin{pmatrix} \sqrt{0} & 0 & 0 \\ 0 & \sqrt{247.16} & 0 \\ 0 & 0 & \sqrt{2.84} \end{pmatrix} \cdot \begin{pmatrix} 0.8498 & -0.5230 & 0.0654 \\ 0.4107 & 0.7349 & 0.5397 \\ -0.3303 & -0.4318 & 0.8393 \end{pmatrix}$$

which should give us matrix A, but it will not! If we do the above multiplication we will not get matrix A. We get something else, something very different that is not correlated with A. The question is why. The answer is that the Singular Value Decomposition has a certain ambiguity with signs of the eigenvectors and eigenvalues.

By the way, from now on, in case we are talking about the Singular Value Decomposition, we will not refer to the eigenvectors any more. We will talk about the left and right singular vectors. ***The left singular vectors*** are those corresponding to matrix U and the ***right singular vectors*** are those corresponding to V. Besides the coefficients of matrix Σ are called the singular values of A.

Now let us talk about these sources of ambiguity. Let us talk about the singular values first. As a convention, the singular values are listed from the biggest to the smallest along the diagonal of Σ. We know that if A is a square full rank matrix, then after applying the Singular Value Decomposition we will get all positive nonzero eigenvalues, that is, nonzero singular values. However, in case A is not full rank, then at least one singular value will be zero. Therefore, as a convention, we will list the singular values from major to minor along the diagonal of Σ.

Another source of ambiguity are the signs of the singular vectors we found, since even if we do the above arrangement about listing the singular values from major to minor, and applying the corresponding changes to matrices U and V, we would still not get matrix A. This is because of the direction of the singular vectors. Let us recall that the left singular vectors are the eigenvectors of $A \cdot A^T$, and that the right singular vectors are the eigenvectors of $A^T \cdot A$. If we scale an eigenvector by any quantity, whether it is positive or negative, the result is still an eigenvector. For example, one of the eigenvectors of $A^T \cdot A$ is the following, but if we change its direction, it will still be an eigenvector:

$$\vec{e_1} = \begin{pmatrix} 0.8498 \\ -0.5230 \\ 0.0654 \end{pmatrix}; -\vec{e_1} = \begin{pmatrix} -0.8498 \\ 0.5230 \\ -0.0654 \end{pmatrix} \rightarrow stillaneigenvector$$

But changing the sign of an eigenvector will also change the result. From here we can notice this source of ambiguity. Many interpretations and procedures are done to solve this ambiguity. Usually, LAPACK softwares do this on their own, but it is important to note that there is no mathematically closed way to solve the problem for the signs or directions of the singular vectors. In case you are more interested in getting more information about this phenomenon, you can look for the paper with the title "Resolving the Sign Ambiguity in the Singular Value Decomposition" wrote by the Sandia National Laboratories.

However, to solve for this ambiguity we will look for an easier way for this process to make sense. We will calculate matrix ***U*** by the following way:

$$U = A \cdot V \cdot \Sigma^{-1}$$

Inverting matrix ***Σ*** is instantaneous since it is a diagonal matrix. However, since ***Σ*** has a zero element in the diagonal, it is not invertible. But this is not a problem at all, because we can cut our matrix to get all the zeros out. This is a usual procedure and it is also done minimize the space in memory that occupies the decomposition of matrix ***A***. Also, we have just stepped into something that is very important with the Singular Value decomposition, and that is the sizes of matrices ***U***, ***Σ*** and ***V***. Assuming ***A*** is an arbitrary matrix of size $m \times n$ and rank ***r***, then ***U*** will be a matrix of size $m \times r$, ***Σ*** will be a matrix of size $r \times r$ and ***V*** will be a matrix of size $r \times n$.

Continuing our example and by doing this arrangements we get:

$$V = \begin{pmatrix} 0.4107 & -0.3303 \\ 0.7349 & -0.4318 \\ 0.5397 & 0.8393 \end{pmatrix}; \Sigma = \begin{pmatrix} \sqrt{247.16} & 0 \\ 0 & \sqrt{2.84} \end{pmatrix}$$

And from here matrix ***U*** is:

$$U = \begin{pmatrix} 0.2226 & 0.7856 \\ 0.7916 & 0.2000 \\ 0.5690 & -0.5856 \end{pmatrix}$$

Therefore, our Singular Value Decomposition is:

$$A = U \cdot \Sigma \cdot V^T = \begin{pmatrix} 0.2226 & 0.7856 \\ 0.7916 & 0.2000 \\ 0.5690 & -0.5856 \end{pmatrix} \cdot \begin{pmatrix} \sqrt{247.16} & 0 \\ 0 & \sqrt{2.84} \end{pmatrix} \cdot \begin{pmatrix} 0.4107 & 0.7349 & 0.5397 \\ -0.3303 & -0.4318 & 0.8393 \end{pmatrix}$$

which now exactly corresponds to the Singular Value Decomposition of ***A***. However, now that we have seen the Singular Value Decomposition of a square matrix, let us talk about its other great advantages -- that the SVD is also possible on rectangular matrices. This allows us to make a decomposition similar to an eigendecomposition on rectangular matrices. Let us look at an example of this to fully understand how it is done. Suppose now we have our matrix ***B*** and we want to compute its Singular Value Decomposition:

$$B = \begin{pmatrix} 1 & 2 & 3 \\ 4 & 5 & 6 \end{pmatrix}$$

At first sight we notice that the rank of ***B*** is 2. This is a valuable information since we know that ***U*** will be a matrix of 2×2. $\boldsymbol{\Sigma}$ will also be a matrix of 2×2, but ***V*** will be a matrix of 2×3. Proceeding with our problem, we now look for the singular values and the right singular vectors (the vectors corresponding to matrix ***V***). Therefore,

$$B^T \cdot B = V \cdot \Sigma^2 \cdot V^T$$

For sure we know that one eigenvalue is zero (***hint***: dependent columns times dependent rows), therefore, we look for the other two:

$$V = \begin{pmatrix} 0.4286 - 0.8059 \\ 0.5662 - 0.1123 \\ 0.7039 \ 0.5812 \end{pmatrix}; \Sigma = \begin{pmatrix} \sqrt{90.4026} & 0 \\ 0 & \sqrt{0.5973} \end{pmatrix}$$

From here we can compute matrix ***U***

$$U = \begin{pmatrix} 0.3863 & 0.9225 \\ 0.9223 & -0.3860 \end{pmatrix}$$

Therefore, our singular value decomposition for matrix ***B*** is

$$V = \begin{pmatrix} 0.4286 - 0.8059 \\ 0.5662 - 0.1123 \\ 0.7039 \ 0.5812 \end{pmatrix}; \Sigma = \begin{pmatrix} \sqrt{90.4026} & 0 \\ 0 & \sqrt{0.5973} \end{pmatrix}$$

Now that we have discussed this, it would be a good time to talk about serious applications of the Singular Value Decomposition. It has applications on face recognition software, data storage, image compression, least squares approximations, checking noise on signal processing, statistics, data analysis, and more. The Singular Value Decomposition has these applications since it is a way of splitting up matrix ***A*** into three pieces, and each piece represents something. Mainly when using these Decompositions, we look at the singular values, since the higher the singular value the greater the corresponding left singular vector affect the matrix and what it represents.

If we take a look again at our last example corresponding to matrix B, we will see that we have two singular values. However, we noticed that the second one is very small in comparison with the first one. Terefore, we can conclude that the second left singular vector does not have a great impact in our matrix.

These kinds of conclusions are made mostly on least squares approximations. For example, if we would be approximating a set of data with a line, then we would only need one singular value. Therefore, if the other singular values are small then they could represent the noise in the system, or the variance in our measurements. However, in case both of our singular values are "big," then we can conclude that approximating our hypothetical data with a line is a bad choice. This was just a hypothetical example of the applications of the SVD, however, it is one of the most easy to understand.

QUADRICS

We already mentioned quadrics at the beginning of this chapter, but we only saw them as a way of getting with the Conjugate Gradient method. However, now we shall see them more deeply. Quadrics are a certain and particular form of surfaces that, in general, come from a polynomial of several variables whose degree is 2. In other words, a quadric, let us call it ***P*** is this:

$$P = \sum_{j=1}^{n} \sum_{i=1}^{n} a_{ij} \cdot x_i \cdot x_j + \sum_{k=1}^{n} b_k \cdot x_k + c$$

wherein we know that the different x_i and x_j represent a variable, and the ***a*** coefficients are any coefficients belonging to a field. This polynomial has very interesting and great properties in mathematics, because it is a generalization of the conical sections like parabolas, circles, ellipses, cones, spheres, ellipsoids and more. These are general surfaces and curves that appear in nature. In fact, many phenomena are either quadratic or linear.

However, sometimes we may encounter problems with these kinds of surfaces. For example, we can easily recognize some of the following surfaces:

$$x^2 + y^2 + z^2 = 1$$

$$\left(\frac{x-h}{a}\right)^2 + \left(\frac{y-k}{b}\right)^2 = 1$$

$$x^2 - z^2 = 1$$

Which correspond to a sphere, an ellipse and a hyperbola, respectively. These are easy to solve and graph quadrics, because they are the common ones. However, in some cases we can have equations of the following way.

$$x^2 + xy + y^2 = 0$$

Where we see there is a mixed term that does not fit quite right with what we know about quadrics. Before showing how to solve these types of equations, let us first return to quadrics and look for help in linear algebra. We have already seen that the above general expression for quadric ***P***, can be factored like this:

$$P = \overrightarrow{x^T} \cdot A \cdot \vec{x} + \overrightarrow{x^T} \cdot \vec{b} + c$$

where ***A*** is a coefficient matrix and ***b*** is a vector containing the coefficients of the single variables, and ***c*** is a constant. Our goal here is to find an algorithm that enables us to graph the above equation. But for this, we need to find a way to handle that mixed term we saw before, and in case we have more than two variables, all of the mixed terms that will arise. For this, let us notice something very intuitive. If ***A*** is a diagonal matrix, then no mixed term appears. Let us look at this in the following equation:

$$\overrightarrow{x^T} \cdot A \cdot \vec{x} = a_{11} \cdot x_1^2 + a_{22} \cdot x_2^2 + \cdots + a_{nn} x_n^2$$

Therefore, to be able to handle a quadric into something we know, we must diagonalize ***A.*** Gladly we saw how to do this in one of the previous chapters. Besides, in this case, we can trick matrix ***A*** to be symmetrical which is advantageous for us. Let us take a look at why we can make matrix ***A*** symmetrical. Suppose we have the following polynomial of two variables.

$$x^2 + 6xy + y^2 = 0$$

If we apply what we just saw then we can factor that polynomial like this:

$$\begin{pmatrix} x & y \end{pmatrix} \cdot \begin{pmatrix} 1 & 3 \\ 3 & 1 \end{pmatrix} \cdot \begin{pmatrix} x \\ y \end{pmatrix} = 0$$

which is symmetrical, however, we could also factor it like this:

$$\begin{pmatrix} x & y \end{pmatrix} \cdot \begin{pmatrix} 1 & 1 \\ 5 & 1 \end{pmatrix} \cdot \begin{pmatrix} x \\ y \end{pmatrix} = 0$$

which is valid, but as we have seen, working with symmetrical matrices is far better than working with other matrices. Recall that in the Conjugate Gradient method we required matrix ***A*** to be symmetrical because we were solving a system of equations with the aid of a quadric. But now we are doing the opposite -- we are solving how to graph a quadric with the aid of a matrix. In reality, if a matrix has any special condition, it can be seen as a blessing since it represents a possibility of a beneficial situation; therefore, let us keep our symmetric arrangement. It is important to notice that this arrangement can be done for any quadric of any number of variables.

Now that we know that we can always factor our quadric with a symmetric matrix, and that we are able to, through an eigendecomposition, diagonalize our matrix, how would we be able to continue from here? The answer to this question is easy. We know that as we will make ***A*** symmetrical, its factorization is the following:

$$A = Q \cdot \Lambda \cdot Q^T$$

If we introduce this into our main equation we get:

$$P = \overrightarrow{x^T} \cdot Q \cdot \Lambda \cdot Q^T \cdot \vec{x} + \overrightarrow{x^T} \cdot \vec{b} + c$$

We can see that we can also make the following arrangements:

$$P = \left(Q^T \cdot \vec{x}\right)^T \cdot \Lambda \cdot Q^T \cdot \vec{x} + \left(Q^T \cdot \vec{x}\right)^T \cdot Q^T \cdot \vec{b} + c$$

Where we can rewrite it as this:

$$Q^T \cdot \vec{x} = \vec{x}\,'$$

$$Q^T \cdot \vec{b} = \vec{b}\,'$$

$$P = \vec{x}\,' \cdot \Lambda \cdot \vec{x}\,' + \vec{x}\,' \cdot \vec{b}\,' + c$$

And from here we can notice three very important things. The first one is that as $\boldsymbol{Q}$ is an orthonormal matrix that transforms our first variables (vector $\boldsymbol{x}$) into a new set of variables (which are contained in vector $\boldsymbol{x'}$). Second, we found a way to diagonalize the coefficient matrix and work with this system easier. And finally, we can get a very interesting geometrical conclusion of this. As matrix $\boldsymbol{Q}$ is an orthonormal matrix we can view it as a matrix that is rotating our initial set of variables into a new set where we do not have any mixed term appearing. This is very desirable because we now have a way to graph and work with any possible multivariate polynomial.

Let us look at some easy examples to practice how this goes. Suppose we have the following polynomial:

$$17 \cdot x^2 - 12xy + 8 \cdot y^2 = 4$$

If we apply the methodology we just explained, we can factor this and get:

$$\begin{pmatrix} x & y \end{pmatrix} \cdot \begin{pmatrix} 17 & -6 \\ -6 & 8 \end{pmatrix} \cdot \begin{pmatrix} x \\ y \end{pmatrix} = 4$$

If we look for the eigendecomposition of the coefficient matrix we would get the following:

$$\lambda_1 = 5; \lambda_2 = 20; \vec{e_1} = \frac{1}{\sqrt{5}} \begin{pmatrix} 1 \\ 2 \end{pmatrix}; \vec{e_2} = \frac{1}{\sqrt{5}} \begin{pmatrix} -2 \\ 1 \end{pmatrix}$$

Where we can factor the coefficient matrix like this:

$$A = \begin{pmatrix} 17 & -6 \\ -6 & 8 \end{pmatrix} = Q \cdot \Lambda \cdot Q^T = \frac{1}{5} \cdot \begin{pmatrix} 1 & -2 \\ 2 & 1 \end{pmatrix} \cdot \begin{pmatrix} 5 & 0 \\ 0 & 20 \end{pmatrix} \cdot \begin{pmatrix} 1 & 2 \\ -2 & 1 \end{pmatrix}$$

And introducing this into our main equation,

$$\begin{pmatrix} x & y \end{pmatrix} \cdot \frac{1}{5} \cdot \begin{pmatrix} 1 & -2 \\ 2 & 1 \end{pmatrix} \cdot \begin{pmatrix} 5 & 0 \\ 0 & 20 \end{pmatrix} \cdot \begin{pmatrix} 1 & 2 \\ -2 & 1 \end{pmatrix} \cdot \begin{pmatrix} x \\ y \end{pmatrix} = 4$$

$$Q^T \cdot \vec{x} = \frac{1}{\sqrt{5}} \cdot \begin{pmatrix} 1 & 2 \\ -2 & 1 \end{pmatrix} \begin{pmatrix} x \\ y \end{pmatrix} = \begin{pmatrix} x' \\ y' \end{pmatrix} = \vec{x}'$$

The one-fifth coefficient can be absorbed by the diagonal matrix, and therefore,

$$\begin{pmatrix} x' & y' \end{pmatrix} \cdot \begin{pmatrix} 1 & 0 \\ 0 & 4 \end{pmatrix} \cdot \begin{pmatrix} x' \\ y' \end{pmatrix} = 4$$

If we expand we will see that our resulting quadric is an ellipse:

$$(x')^2 + 4 \cdot (y')^2 = 4$$

From the last result we know we have an ellipse in terms of our new set of variables, but now we also need to know how our old set of variables is represented with respect to the new ones to be able to graph this ellipse properly. This is done easily by looking at the following relationship:

$$\vec{x}' = Q^T \cdot \vec{x} \rightarrow \vec{x} = Q \cdot \vec{x}'$$

We can see that the column vectors in our matrix $\boldsymbol{Q}$ form the new set of Coordinate axes to graph our ellipse. This can be seen in the following image.

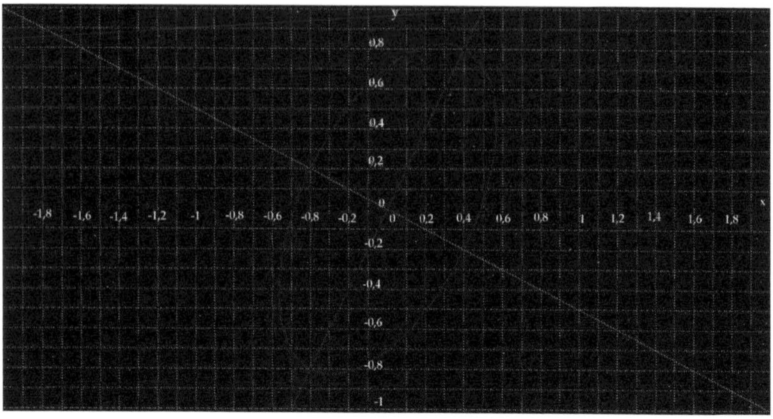

Figure 25: Graph of our ellipse.

We can see in Figure 23 an ellipse rotated by a certain angle (which can be calculated from matrix $\boldsymbol{Q}$).

CHAPTER 8

POLYNOMIALS

"Polynomials in one variable, in ten variables, in twenty seven variables, who cares? (long pause) As long as it's finite."

Abstract algebra professor

INTRODUCTION

Polynomials are a central part of linear algebra. We touched this subject briefly in past chapters while talking about quadrics, but because it was not the main theme, we did not have an in depth study of them. Therefore, this chapter is especially dedicated to polynomials. Polynomials are functions consisting of variables, coefficients, and non-negative integer exponents of variables, and involve only the operations of addition, subtraction, multiplication, and division.

In past chapters, we mentioned quadrics, and we mentioned that polynomials can be univariate or multivariate, of a certain degree. We saw that a quadric is a polynomial of degree two, and it does not matter how many variables it has. However, regular polynomials can be of any degree. In polynomials any combination is possible.

In this chapter we will focus on the methods to solve for the roots of univariate polynomials. We will also give attention to multivariate polynomials, but not as much as univariate polynomials, since they have more theoretical applications than practical. In this chapter we will also see the importance of numerical methods, and we will see that it is relatively easy to solve for the roots of univariate polynomials with numerical methods like Laguerre's method, Halley's method, Muller's method and Newton's method.

Also, we will talk about how to allocate each zero of a polynomial like we did with eigenvalues with the Gershgorin theorem. Finally, we will also talk about multiplicities of roots and how they affect our methods.

POLYNOMIALS

Polynomials are an essential part of algebra, not only of linear algebra. That is the reason why we are including this chapter. It is important to know about the history of polynomials before going deep into them, since polynomials, as matrices, have been wandering the earth from millenniums ago. For example, in ancient Babylonia it was a common knowledge to solve for quadratic equations of the form:

$$x^2 + qx + p = 0$$

Where $\boldsymbol{q}$ and $\boldsymbol{p}$ represents any number. Later on, during the seventh century, Muslim mathematicians expanded knowledge even to China in solving cubic equations and quartic equations. With the arrival of the dark ages, most knowledge stagnated, but after the renaissance, mathematics knowledge was reborn. It became a common knowledge to solve for cubic equations. German scholars advanced in trigonometry, Greeks advanced in astronomy and physics, and the first Europe's great mathematician was known, Leonardo of Pisa, better known as Fibonacci, who is famous for his Fibonacci numbers, however, that is not his only nor his greatest contribution to mathematics.

Polynomials, in this sense, is always present in real life situations, because they are able to model many phenomena. You may have already known that polynomials are a certain type of function that are most desired because of their roots. Let us talk about this at first with univariate polynomials. Univariate polynomials are functions of this form:

$P(x) = a_0 + a_1x + a_2x^2 + ... + a_nx^n$

We can see that the above function only uses addition, subtraction, multiplication, division and positive integer exponent of variables. A function takes some value ***x*** and returns some value ***P(x)***. It is easy to evaluate this function at a certain point ***x*** having the above function, however, it can be immensely difficult to do the opposite, that is, look for the value of ***x*** that gives a certain value ***P(x)***. This last situation has been cracking mathematicians' heads through centuries. Let us talk about this problem extensively. The above general expression can be rewritten like this:

$(a_0 - P(x)) + a_1x + a_2x^2 + ... + a_nx^n = 0$

Therefore, depending on the value of ***P(x)*** we will have a certain set of solutions of the above polynomial. But let us focus on the real idea behind this. Young mathematicians are usually taught this expression at the beginning of their formation, $ax - b = 0$, which is a linear single variable equation. However, we have seen this case extensively in its multivariate form. Continuing with our history, mathematicians then faced harder problems, and at some point were presented with this: $ax^2 + bx + c = 0$, which is the regular quadratic equation. It has a very famous solution, which is

$$x = \frac{-b \pm \sqrt{b^2 - 4 \cdot a \cdot c}}{2a}$$

According to researchers, Babylonians knew how to solve for quadratic equations by reducing them to their normal form, followed by a method for solving the normal form. The normal form involves a task of finding two numbers given their sum and their product. In modern algebraic notation, this can be stated like this: Given two numbers ***p*** and ***s***, and given that $xy = p$ and that $x + y = s$ find ***x*** and ***y***.

The steps by which the Babylonians solved this problem are the following:

1. Take half of ***s***.
2. Square the result.

3. From this, subtract ***p***.
4. Take the square root of the result.
5. Add this to half of ***s***; this is one of the two numbers and the other is ***s*** minus this one.

If we follow these steps we will reach the usual quadratic formula for when $a=1$. In many modern schools, ways to analytically solve for polynomials of order higher than 2 are not taught, but that does not mean that they do not exist. Actually generic equations to solve for polynomials exist for polynomials of at most degree 4, since it is impossible to solve analytically for polynomials of degree 5 or more. The problem of solving the quintic equation troubled many mathematicians for centuries, including Gauss. However, it was not until 1824 that Niels Able gave a complete proof of the impossibility of solving the quintic equation. This is called as the Able-Ruffini Theorem, the proof of which is usually demonstrated through Galois Theory. We will not demonstrate this theorem, however. The next subject we are going to see is a very important theorem in mathematics, which is the Fundamental Theorem of Algebra.

FUNDAMENTAL THEOREM OF ALGEBRA

The Fundamental Theorem of Algebra states that "*an* $\boldsymbol{n}^{th}$ *degree polynomial has* ***n*** *roots, whether they are all equal or distinct, real or complex.*" This comes from the fact that every polynomial can be factored in terms of its roots, since a polynomial of degree 2 can only be formed by the product of two polynomials of degree 1. This theorem is important since it gives important ways to solve certain problem, for example, the Basel problem which will be discussed further in this chapter.

We mentioned the word roots. You should already know what a root is, but just in case you don't know, a ***root***, also called a zero, is a value of ***x*** such that $P(x)=0$. Looking for the roots of a polynomial can be a hard problem, but there are certain rules and theorems that are useful for allocating these roots. For this reason we will look at a certain set of bounds that may give us useful information to be able to compute for the roots of polynomials. The first one of them is ***Descartes' Rule of Signs***, which is an easy way to check for an upper bound for the number of positive and negative real roots of a polynomial.

Bounds on Roots of Polynomials

Descartes rule of signs says that, "*for a single-variable polynomial with real coefficients, ordered by descending variable exponent, the number of positive roots is either equal for every change of signs in consecutive coefficients of a polynomial or less than it by an even number.*" There is also a corollary of this rule, which corresponds by making the change of variable of x by $-x$, which gives the number of negative roots of the polynomial.

There is also a known fact about polynomials that says that a certain polynomial ***p(x)*** has at least one root inside the circle of radius ρ_1 about the origin, the radius given by the following expression:

$$\rho_1 = min\left[n \cdot \left| \frac{a_0}{a_1} \right|, \sqrt[n]{\left| \frac{a_0}{a_n} \right|} \right]$$

Where ***n*** is the degree of the polynomial, and a_i are the different coefficients of the polynomial. Another bound on the roots polynomial says that all zeros of a certain polynomial ***p(x)*** are all inside a circle of radius ρ_2 about the origin, being the radius given by the following expression:

$$\rho_2 = 1 + max\left| \frac{a_i}{a_n} \right| \rightarrow i = 0,1,2,...,n-1$$

These last two methods may seem small facts about polynomials, however, they have been known for centuries and are a way of looking at where the zeros of a polynomial might be. A more complete method for finding the roots of a certain polynomial is given by ***Sturm's Theorem***, which says that "*the number of distinct real zeros of a polynomial **f(x)**, with real coefficients, in a certain interval **(a, b)** is equal to the excess of the number of changes of signs in the sequence* $f_0(a),...,f_{k-1}(a),f_k(a)$ *over the number of changes of sign in the sequence* $f_0(b),...,f_{k-1}(b),f_k(b)$, *where the different functions* $f_0, f_1,..., f_n$ *are given by the following:*

$$f_0 = p(x) \qquad f_3 = -rem(f_1, f_2)$$

$$f_1 = p'(x) \qquad f_k = -rem(f_{k-2}, f_{k-1})$$

$$f_2 = -rem(f_0, f_1) \quad f_{n+1} = -rem(f_{n-1}, f_n) = 0$$

where $rem(f_i, f_j)$ *is the reminder, or quotient, of the division of the polynomials* f_i *and* f_j." You might be confused with what you just saw, but there is no need to worry as all will be cleared out soon. Let us keep in mind that this theorem will be very useful and simple once we get the hang of it.

Before getting deeper into roots, let us talk more about Sturm's theorem. Sturm's Theorem was presented during the nineteenth century by Jacques Charles Sturm as a way of allocating the roots of a polynomial. The thing is that Sturm's theorem is able to tell us how many real roots of a polynomial are at a certain interval. It does this by the sign changes of the consecutive functions we showed before. As a last comment on Sturm's theorem for now, suppose the sign changes are denoted by a number sigma which will depend on the interval ***(a,b)***. Therefore, the number of real roots between the interval ***a*** and ***b*** is:

$N°\,roots = \sigma(a) - \sigma(b)$

Now would be a good time to see these methods in action. It is understandable if you feel confused about what we just saw, but there is no need to worry. These methods are very straightforward and simple. To explain these methods with higher detail let us make an example. Suppose we have the following polynomial ***p(x)***, and we want to find its roots or at least find the intervals in which those roots exist:

$p(x) = x^4 - 3x^3 + x^2 + 2x - 4$

First, let us try and get information easily by Descartes' rule of signs. We can see that we have the polynomial's coefficients arranged from highest degree to lowest degree (from degree 4 to the independent term). Descartes' rule of signs says that the number of positive roots of the polynomial is equal to the number of change of signs of consecutive coefficients of the polynomial or less than it by an even number. In our example we see that the coefficients have positive, negative, positive, positive and negative sign respectively. Therefore, we see we have three change of signs, which means that we have either 3 or 1 positive real roots.

Now we can also get information on the number of negative roots, by making the change of variable $x \to -x$. If we do this, we get the following:

$$p(x) = x^4 - 3x^3 + x^2 + 2x - 4$$

Where we see that the consecutive coefficients have positive, positive, positive, negative and negative sign respectively. Meaning we only have a change of sign, which implies that we only have a single negative real root. Now let us try and find other information with the other two bounds we saw before. The first radius ρ_1 is given by this expression:

$$\rho_1 = min\left[n \cdot \left| \frac{a_0}{a_1} \right|, \sqrt[n]{\left| \frac{a_0}{a_n} \right|} \right]$$

$$\rho_1 = min\left[8, \sqrt{2} \right] = \sqrt{2}$$

And the second radius ρ_2 is given by

$$\rho_2 = 1 + max\left| \frac{a_i}{a_n} \right| \to i = 0, 1, 2, \ldots, n-1$$

$$\rho_2 = 1 + max\left| \frac{-3, 1, 2, -4}{1} \right| = 1 + \left| \frac{-4}{1} \right| = 5$$

Therefore, we now know that at least one root is within a circle of radius $\sqrt{2}$ with center above the origin, and that all the roots are contained within a circle of radius 5 with center above the origin. This is one of the easiest ways to see where the roots of our polynomial might be. Notice that our polynomial is of degree 4. Therefore, we have 4 roots, we are sure that we have 1 real negative root, we know that we might have 3 or

1 real positive roots, and that in case we only have one positive real root, then we would have 2 complex roots.

Now let us see how to use Sturm's theorem. Sturm's theorem is more complicated than this, however, but it is still a very good method to check which intervals we are interested in to see if there are roots. Besides, it also gives higher detail than the methods we just saw. Sturm's theorem is formed by a certain set of functions which we expressed how to calculate before, however, you might be confused with the actual computation of these functions. Recall what we said about the first two functions, that they are the polynomial and its derivative:

$$f_0 = p(x) = x^4 - 3x^3 + x^2 + 2x - 4$$

$$f_1 = p'(x) = 4x^3 - 9x^2 + 2x + 2$$

And that the following functions are the negative reminders of the division of consecutive functions. If we want to calculate the second function, for example, we do the following:

$$\frac{f_0}{f_1} = \frac{p(x)}{p'(x)} = \frac{x}{4} - \frac{3}{16} + \frac{-19x^2 + 30x - 58}{16 \cdot (4x^3 - 9x^2 + 2x + 2)} \rightarrow rem(f_0, f_1) = \frac{-19x^2 + 30x - 58}{16}$$

And $\boldsymbol{f_2}$ is given by:

$$f_2 = -rem(f_0, f_1) = \frac{19x^2 - 30x + 58}{16}$$

If we do this approach with consecutive functions we would get:

$$f_0 = x^4 - 3x^3 + x^2 + 2x - 4$$

$$f_1 = 4x^3 - 9x^2 + 2x + 2$$

$$f_2 = \frac{19x^2 - 30x + 58}{16}$$

$$f_3 = \frac{83456x - 58880}{361}$$

$$f_4 = -\frac{1229927}{425104}$$

which is a set of five functions, and it will tell us how many zeros there is between a certain range of values. For this we need to evaluate these functions in that range. Suppose we want to evaluate them between 0 and 5. If we do this we get:

Table 1: Evaluated functions of the Sturm functions

Function	$x = 0$	$x = 5$
f_0	-4	281
f_1	2	287
f_2	58/16	383/16
f_3	-58880/361	358400/361
f_4	-1229927/425104	-1229927/425104

And with the above chart, we do a procedure equivalent to Descartes' rule of signs. We seek for the sign changes at both sides, at x=0 and at x=5. We see that at x equal to zero we have two sign changes and at x equal to five we have only one sign change. In this case we have that the number of real roots in this interval is:

$$\sigma(0) = 2; \sigma(5) = 1 \rightarrow N^{\circ}\, roots(0,5) = \sigma(0) - \sigma(5) = 1$$

It would be nice to check other properties which would make the computation for Sturm's theorem easier. Let us look at what this theorem really does.

While applying Sturm's theorem we calculate a certain set of functions (which are commonly called a ***Sturm chain***), we evaluate them at a certain interval, and look at the sign changes of those functions at that interval. That is the key -- we are not interested at the values of those functions, we are only interested in their signs, and this allows us to make things easier.

We can make Sturm's theorem easier by multiplying each function in the Sturm chain by a certain, nonnegative and different from zero, constant that makes it easier to evaluate them and also to divide the consecutive polynomials. Let us make this clear to understand, since we are only interested in the signs of the functions of the Sturm chain, then we can multiply a function of the chain by a positive constant without changing the sign of that function at any point.

Going back to our preceding Sturm chain and multiply each function by a certain constant to make it easier to evaluate at a certain interval, we would get:

$$f_0 = x^4 - 3x^3 + x^2 + 2x - 4$$

$$f_1 = 4x^3 - 9x^2 + 2x + 2 \qquad f_3 = 163x - 115$$

$$f_2 = 19x^2 - 30x + 58 \qquad f_4 = -1$$

Now that we have this, let us make our interval smaller so we can get a better look at the root we are looking for, let us say our new interval will be between 1 and 3.

Table 2: Evaluated functions of the simplified Sturm chain

Function	$x = 1$	$x = 3$
f_0	-3	11
f_1	-1	89
f_2	47	139
f_3	48	374
f_4	-1	-1

$$\sigma(1) = 2; \sigma(3) = 1 \rightarrow N^\circ roots(1,3) = 1$$

Therefore, we now have reduced our search interval and may apply any root finding algorithm the reader knows to look for the root, however, Sturm's theorem can be used between several intervals to check for distinct roots. Since we have mentioned this, it would be a very good time to check for some root finding algorithms.

ROOT FINDING ALGORITHMS

You may probably be aware of some root finding algorithms, mainly the ***bisection method***, ***Regula Falsi***, ***Newton-Raphson***, ***quasi Newton methods***, and ***Muller's method***. For this reason, we will only discuss briefly about the usefulness of these methods but not their properties nor how these methods actually work. It is important to remember that these methods were developed some decades ago, and decades ago humanity has not faced problems or situations like it does now.

The idea behind the last sentence is that, it is true that the methods we just mentioned before may be slower than others, sometimes they may fail and, even sometimes, they may have many errors in their results, but that is because they were born and used in a simpler world with not so many requirements. However, they are still shown today because they are simple to use, they have their particular advantages and can be modified to give better results than their unrefined ancestors. Let us give a few comments on those methods we mentioned and explain Muller's method and other faster and newer methods that are easy to implement, have high rate of convergence and are quite difficult to diverge.

Bisection

Bisection is a method that is easy to understand. It is one of the few methods that is assured to converge to the solution of a root within a certain range of error. However, as bisection tends to be a linear root finding algorithm it is not used quite often nowadays, sadly because many improvements can be done to the original method.

Regula Falsi and Secant

Regula Falsi is also a linear root finding algorithm that always converges to a solution of a polynomial by means of interpolating our polynomial with a linear approximation. Usually, Regula Falsi converges faster than bisection, however, sometimes it works even slower. One case of this would be with functions which has very high derivative along the interval we are working in. Meanwhile, Secant method is a certain of step between Regula Falsi and Newton-Raphson method since it relates to both. Because of this, the Secant Method is faster than Regula Falsi but it is not always assured to converge.

Newton-Raphson Method

This method, also known as Newton's method, is a simple algorithm to find roots of a polynomial. It has order of convergence 2, in case it is after a root of multiplicity 1. It was developed by Isaac Newton and Joseph Raphson in parallel almost at the same time, however, Raphson managed to publish it before Newton did. This method is one of the most used today because of its relatively high order of convergence and simplicity, however, Newton's method is not a safe method because it diverges very easily and sometimes can get upon trouble if it gets very close to a maximum or a minimum of the polynomials we are working on.

Muller's Method

Muller's method is also an iterative root finding algorithm which interpolates a polynomial with a quadratic approximation. This quality gives it speed of convergence higher than one. Muller's method is a little bit slower than Newton's method in most cases, but it presents an advantage that other methods do not. It is able to give complex roots for a polynomial given a starting real seed, which means that it is a very useful tool to get in touch with those complex roots that are omitted by other methods. In addition, Muller's method does not require the computation of any derivative.

Muller's method, as we have said before, is an iterative method that approximates roots of a polynomial. It does so by approximating the root through a set of parabolas formed from three consecutive points and calculating the next point that the parabola intersects in the x axis. Usually, some authors start with only two points and calculate a third one with the Secant method or Regula Falsi, but this is not necessary. A detail about this method is that a parabola is a quadric that intersects the x axis in two different points, but to assure convergence what is done is to take the solution closest to the last iterated point.

This method was presented by ***David Muller***, a former American mathematician and computer scientist who has also other advancements in different areas like electronics and Group theory. This method was presented in 1956. Geometrically, Muller's method looks like this:

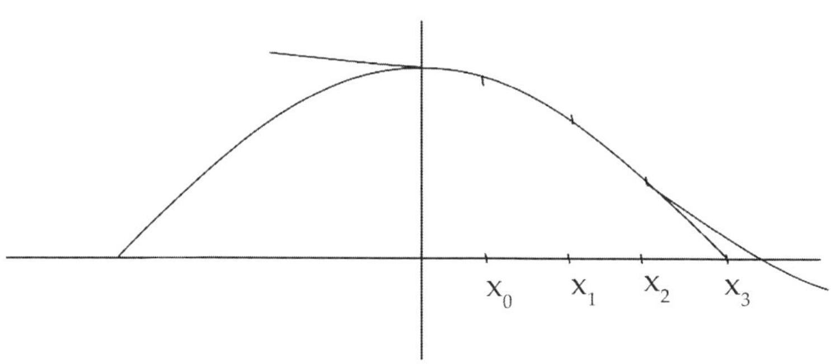

Figure 26: Muller's method.

We can see in the graph above that we have three initial points x_0, x_1, and x_2 which are used to form a parabola. This parabola approximates the function we are studying through a certain range. Therefore, the roots of our parabola approximate those of our function. To do this, assuming we have the three initial points mentioned before with their respective images y_0, y_1, and y_2. We would form a parabola with the following equation:

$$p(x) = a \cdot (x - x_2)^2 + b \cdot (x - x_2) + y_2$$

where ***a*** and ***b*** are given by:

$$a = \frac{(x_1 - x_2) \cdot (y_0 - y_2) - (x_0 - x_2) \cdot (y_1 - y_2)}{(x_0 - x_2) \cdot (x_0 - x_1) \cdot (x_1 - x_2)}$$

$$b = \frac{(x_0 - x_2)^2 \cdot (y_1 - y_2) - (x_1 - x_2)^2 \cdot (y_0 - y_2)}{(x_0 - x_2) \cdot (x_0 - x_1) \cdot (x_1 - x_2)}$$

If we want to look for the point where our parabola intersects the ***x*** axis, we will set ***p(x)*** equal to zero and solve for ***x***. However, in this case, to avoid catastrophic cancellation, what is done is to do the following arrangement:

$$a \cdot (x - x_2)^2 + b \cdot (x - x_2) + y_2 = 0$$

$$\frac{y_2}{(x - x_2)^2} + \frac{b}{(x - x_2)} + a = 0$$

Solving for ***x*** we get

$$x - x_2 = \frac{-2 \cdot y_2}{b \mp \sqrt{b^2 - 4a \cdot y_k}}$$

And if we would apply this iterative method, we would get

$$x_{k+1} = x_k + \frac{-2 \cdot y_k}{b \mp \sqrt{b^2 - 4a \cdot y_k}}$$

where the sign ambiguity in the denominator is chosen to make the denominator with greater norm in order to achieve convergence. Notice that we may get negative square roots, meaning that this method is able to give complex roots as a result.

A very important detail about complex roots, that the reader may also know, is that in a polynomial with all real coefficients the complex roots come in conjugate pairs. That is, once we found one root we can get the other. Even though Muller's method is a very good approach to solve for roots of a polynomial, it is not the only method available. Another one is ***Laguerre's method***.

Laguerre's Method

Laguerre's method is another numerical method for solving root problems with polynomials. It was proposed by French mathematician ***Edmond Laguerre*** who is most famous for his other contributions in algebra. This method is not very popular nowadays, however, but it is one of the fastest numerical methods that is known with a rate of convergence equal to 3 in case it is after a root with multiplicity equal to 1. In case it is after a root of higher multiplicity, then its rate of convergence is linear. This is a very nice advantage because in many cases it is rare to have a polynomial with roots of high multiplicity.

The method is very simple to demonstrate, however, its proof will also help us with other topics later on. To prove this method, first we need to see a fact about the Fundamental Theorem of Algebra, which is "*since an **n**th degree polynomial has **n** roots, then we can factor that polynomial in terms of its roots.*"

Therefore, if we have a generic polynomial like this:

$$P(x) = a_0 + a_1 x + a_2 x^2 + \ldots + a_n x^n$$

And assuming that our roots are called x_1, x_2, …, and x_n then our polynomial can be expressed like this:

$$P(x) = c(x - x_1)\cdot(x - x_2)\cdot\ldots\cdot(x - x_n)$$

Where c is a constant. The formulation for Laguerre's method comes if we apply the natural logarithm to the above equation. By properties of logarithms we can separate it like the following:

$$\ln(P(x)) = \ln(c) + \ln(x - x_1) + \ln(x - x_2) + \ldots + \ln(x - x_n)$$

If we derive the above expression with respect to x we would get:

$$\frac{P'(x)}{P(x)} = \frac{1}{x - x_1} + \frac{1}{x - x_2} + \ldots + \frac{1}{x - x_n}$$

And if we derive it again we would also get that:

$$\frac{P''(x)}{P(x)} - \left(\frac{P'(x)}{P(x)}\right)^2 = -\frac{1}{(x - x_1)^2} - \frac{1}{(x - x_2)^2} - \ldots - \frac{1}{(x - x_n)^2}$$

Now let us make what is called a drastic set of assumptions. Let us say that we are interested in a root x_i that is at a distance a from our initial guess x, and that all the other roots are clustered together at a certain distance b away from our initial guess x. Therefore,

$$x - x_i = a \quad x - x_j = b \rightarrow j \neq i$$

If we input this conditions in our above equations we would get:

$$\frac{P'(x)}{P(x)} = \frac{1}{a} + \frac{n-1}{b} \qquad \frac{P''(x)}{P(x)} - \left(\frac{P'(x)}{P(x)}\right)^2 = -\frac{1}{a^2} - \frac{n-1}{b^2}$$

To make things easier, it would be nice to pack the above left hand sides of the equations by giving them names, like the following:

$$\frac{P'(x)}{P(x)} = G \quad \left(\frac{P'(x)}{P(x)}\right)^2 - \frac{P''(x)}{P(x)} = H$$

Now that we have this set of equations, we are able to solve for a which is the distance from our guess x to the hypothetical root we are looking for. We will show both our equations in terms of G and H because it is important to notice the signs of the different terms. An error in a sign of an equation may cause the derivation to fail, therefore,

$$G = \frac{1}{a} + \frac{n-1}{b} \quad H = \frac{1}{a^2} + \frac{n-1}{b^2}$$

If we solve for a in the above equations we would get that a is:

$$a = \frac{n}{G \pm \sqrt{(n-1)\left(n \cdot H - G^2\right)}}$$

And our recurrence formula is

$$x_{k+1} = x_k - a$$

Note that we have a sign ambiguity in the calculation of ***a***. This is resolved by choosing the sign that makes the denominator with the greatest norm. Laguerre's method is probably one of the simplest method, in terms of calculation, to search for the roots of a polynomial since it only requires to evaluate our polynomial and its following two derivatives at a certain value ***x***. Doing this by Horner's scheme of Ruffini's rule is an easy task. Laguerre's method also has cubical convergence which is of much benefit since it requires less iterations than other methods to get a reasonable value of the solution. Besides, this method is considered as a sure-fire method because it is extremely rare to diverge while applying it. Also, note that because it involves a square root, this method also gives the possibility to encounter complex roots of a polynomial. However, this method is not the only one with cubical convergence. There is another method that has cubical divergence and it is popularly known as ***Halley's method***.

Halley's Method

Halley's method is another root finding algorithm with cubical rate of convergence. It was invented and presented by English mathematician and astronomer ***Edmond Halley***, which is most famous for computing the orbit of Halley's comet. This method is derived from Newton's method, however, even though its derivation uses another algorithm. These methods were not born far away in time. Actually, Isaac Newton and Edmond Halley lived in the same century and met each other to discuss planetary motions since both were interested in these subjects.

To get the derivation for this method we need to remember Newton's approach to calculate roots.

$$x_{k+1} = x_k - \frac{f(x_k)}{f'(x_k)}$$

Now assume the ***f*** has the following form:

$$f(x) = \frac{g(x)}{\sqrt{g'(x)}}$$

Part of the assumption is that, obviously, ***g*** has the same zeros as ***f***. However, we must also assume that the zeros are of ***g*** are not the ones of its derivative, which means that function ***g*** has no repeated roots or roots of high multiplicity. By assuming these conditions, if we input ***f*** in terms of ***g*** we would get the following expression:

$$x_{k+1} = x_k - \frac{2 \cdot g(x_k) \cdot g'(x_k)}{2 \cdot (g'(x_k))^2 - g(x_k) \cdot g''(x_k)}$$

Notice that this method also requires the computation of our polynomial and its two consecutive derivatives, which means equivalent operations to those of Laguerre's method. However, computing the derivative of a polynomial is fairly easy and with the aid of computation this becomes just a few lines of code.

Now that we have seen this methods it would be a good time to see them in action with a real example. We will show how to compute one iteration with each method (including Newton's method) with the same initial seed value, and then we will just show our results after each iteration, because what we are really interested in, is in the results and how each method behaves. We are also doing this to be able to compare the four methods: ***Newton's***, ***Muller's***, ***Laguerre's*** and ***Halley's***. It is important to say that there exists the possibility that not all methods will converge to the same root with the same seed, so let us keep that in mind.

It is recommended to use a program so you can easily program, or at least evaluate, the functions we are going to use. You can use Scilab, Matlab, FORTRAN, or some online application, but for now, let us look at our example. Let us use the one we had before:

$$p(x) = x^4 - 3x^3 + x^2 + 2x - 4$$

From the analysis we did before we know that this polynomial has one root with absolute value smaller than $\sqrt{2}$, all roots with absolute value less than 5, it has 1 real root contained in the interval (1, 3), and it also has 1 negative real root. Let us apply the methods with a seed value equal to 1.

We will show the first and second derivatives of our polynomial and evaluate them at the initial seed point.

$$p'(x) = 4\cdot x^3 - 9\cdot x^2 + 2\cdot x + 2 p''(x) = 12\cdot x^2 - 18\cdot x + 2$$

$$x_0 = 1 \rightarrow p(x_0) = -3; p'(x_0) = -1; p''(x) = -4$$

- ***Newton's method*** gives as first iteration:

$$x_1 = x_0 - \frac{f(x_0)}{f'(x_0)} = 1 - \left(\frac{-3}{-1}\right) = -2$$

- ***Laguerre's method*** gives as first iteration:

$$\frac{P'(x_0)}{P(x_0)} = G_0 = 0.3333 \quad \left(\frac{P'(x_0)}{P(x_0)}\right)^2 - \frac{P''(x_0)}{P(x_0)} = H_0 = -1.2222$$

$$a_0 = \frac{4}{G_0 \pm \sqrt{3(4\cdot H - G_0^2)}} = 0.9710 \qquad \rightarrow x_1 = x_0 - a = 1 - 0.9710 = 0.0289$$

- ***Halley's method*** gives as first iteration:

$$x_1 = x_0 - \frac{2\cdot f(x_0)\cdot f'(x_0)}{2\cdot(f'(x_0))^2 - f(x_0)\cdot f''(x_0)} = 1 - (-0.6) = 1.6$$

- For ***Muller's method*** remember that we need three initial points (or only 2 initial points and a third one provided from another method like Secant or Regula Falsi). In this case, we will start with three points which are the ones shown below. Muller's method gives as first iteration:

$$x_0 = 0; x_1 = 1; x_2 = 2 \quad y_0 = -4; y_1 = -3; y_2 = -4$$

$$a_0 = -1; b_0 = -2 x_3 = x_2 + \frac{-2\cdot y_2}{b \mp \sqrt{b^2 - 4a\cdot y_k}} = 1 + 1.732\cdot i$$

Now let us look at the following table which demonstrates the iterations for all methods:

Table 3: Comparison of the different root finding iterative methods

Iteration/ Method	*Newton's*	*Muller's*	*Laguerre's*	*Halley's*
1	-2	$1 + 1.732 \cdot i$	0.0289	1.6
2	-1.4857	$1 + 0.5773 \cdot i$	$0.9117 + 1.0252 \cdot i$	1.0160
3	-1.1927	$0.8947 + 0.8204 \cdot i$	$0.7650 + 0.9416 \cdot i$	1.6910
4	-1.0850	$0.7874 + 0.9252 \cdot i$	$0.7650 + 0.9409 \cdot i$	1.3740
5	-1.0708	$0.7643 + 0.9403 \cdot i$		-0.6238
6	-1.0706	$0.7650 + 0.9409 \cdot i$		-0.9232
7				-1.0672
8				-1.0706

We can see from the above table several things. The first one of them is that ***Newton's method*** behaved well for the initial seed value since it managed to calculate one root in only 6 iterations (notice that this is the negative root we were talking about before). ***Muller's method*** also behaved very well by providing a complex root in 6 iterations (remember that complex roots in a polynomial with real coefficients come in conjugate pairs, therefore, we have another root from this approach). ***Laguerre's method*** was the fastest method in computing a root, it only took 4 iterations, which is truly amazing. And ***Halley's method*** was the method that took more iterations to calculate a root, even though it has cubical rate of convergence, the reason why it happened is very interesting.

The reason that Halley's method took more iterations to calculate a root even when it should have converged faster than Newton's method, is the initial seed value we took. At the first iterations of Halley's method we can see that our values of $\boldsymbol{x}$ increased and then suddenly dropped down to the same root computed by Newton's method. This is because our initial seed value for Halley's method is not very good. Actually, during the first iterations, Halley's method was looking for another root until it finally got what it was looking for.

Going back to our example, having used these methods with the same polynomial we obtained three roots, which are:

$$x_1 \approx -1.0706; x_2 \approx 0.7650 + 0.9409 \cdot i; x_3 \approx 0.7650 - 0.9409 \cdot i$$

We are still in need of another root since we must have four roots to complete the problem (which, in fact, is the real positive root that is between 1 and 3). In this instance we could apply one of the methods again with another seed value (for example, Halley's method converges in four iterations to the missing root with a seed value of 2). However, this is not very efficient, because it kind of depends on luck when choosing the next seed value that we might end up getting a root we already calculated before. What is usually done in this instance is to extract the calculated roots from the original polynomial so we don't calculate them again. This is done by factoring the polynomial. In our case, if we first factor out the root we would get:

$$p(x)=(x+1.0706)\cdot\left(x^3-4.0706\cdot x^2+5.35798\cdot x-3.73626\right)+0.0000378$$

wherein we have extracted our root from the original polynomial, but we have an independent term that survived during the factorization. However, we noticed that this term is close to zero, which is good (ideally it should be zero). This term represents our residual, and is a measure of how accurate the calculation of our root is -- the better our root, the smaller that term will be. In this case we can ignore that residual since it is small. Therefore,

$$p(x)\approx(x+1.0706)\cdot\left(x^3-4.0706\cdot x^2+5.35798\cdot x-3.73626\right)$$

The polynomial of third degree that is left on the right side of the above equation is called the deflated polynomial, and it contains all the other roots that we are looking for. Therefore, what is usually done is that, after calculating a root from a polynomial we extract that root and apply some method to the deflated polynomial, thereby looking for the missing roots. In our past example we have three roots, and if we factor all of them out we would get:

$$p(x)\approx(x+1.0706)\cdot\left(x^2-1.53\cdot x+1.47051\right)\cdot(x-2.5404)$$

where the quadratic term corresponds to both of the complex roots. This is the reason why polynomials with real coefficients, if they have complex roots, then they must come in conjugate pairs, because when we multiply those conjugate complex roots the result is a second degree polynomial with real coefficients. As a last comment on this example, we noticed that the last root is 2.5404. Now that we have shown these four methods, which are truly magnificent, it would be nice to take a brief trip over the field of ***symmetric polynomials*** to explain other details about polynomials. As a final comment on the section of iterative methods, the best method according to different authors, that is being used nowadays to compute for roots of polynomials is the Jenkins and Traub algorithm.

SYMMETRIC POLYNOMIALS

Isaac Newton (1643-1727) is most famous for his discovery of the universal law of gravitation and for his use of that law to give an exact mathematical description of planetary motion. However, Newton also had many things to share during his career. For example, he is known with Leibniz as the father of differential calculus and published many papers relating to linear algebra. Newton discovered something very interesting regarding polynomials, and this is the part of the chapter where we begin talking about multivariate polynomials.

Let us talk about what Newton did. He found a way of expressing the roots of a polynomial in terms of the coefficients of that polynomial. This is difficult to see with a general ***n***-term polynomial, so let us do it with a general polynomial of degree 4, like this:

$$p(x) = a_0 + a_1 \cdot x + a_2 \cdot x^2 + a_3 \cdot x^3 + a_4 \cdot x^4$$

In case we would want to get the roots of the above polynomial, we would set it equal to zero. In case we do that, we can divide by $\boldsymbol{a_4}$ without loss of generality to get the following:

$$b_0 + b_1 \cdot x + b_2 \cdot x^2 + b_3 \cdot x^3 + x^4 = 0 \rightarrow b_i = \frac{a_i}{a_4}$$

We know that the above equation has 4 roots which can be unique or repeated, real or complex. Let us call our different roots by the names ***r***, ***s***, ***t***, and ***u***. The above polynomial then can be rewritten as:

$$(x-r) \cdot (x-s) \cdot (x-t) \cdot (x-u) = 0$$

The thing that newton noticed was that if we expand the above polynomial, then we should get the original polynomial we had. But by doing this, he found a way to relate the coefficients of the original polynomial with the roots of that polynomial. Let us look at this by expanding our last expression:

$$x^4-(r+s+t+u)\cdot x^3+(r\cdot s+r\cdot t+r\cdot u+s\cdot t+s\cdot u+t\cdot u)\cdot x^2-(r\cdot s\cdot t+r\cdot s\cdot u+r\cdot t\cdot u+s\cdot t\cdot u)\cdot x+(r\cdot s\cdot t\cdot u)=0$$

wherein we can do the following equalities because both polynomials must be the same:

$$b_3=-(r+s+t+u)$$

$$b_2=(r\cdot s)+(r\cdot t)+(r\cdot u)+(s\cdot t)+(s\cdot u)+(t\cdot u)$$

$$b_1=-[(r\cdot s\cdot t)+(r\cdot s\cdot u)+(r\cdot t\cdot u)+(s\cdot t\cdot u)]$$

$$b_0=r\cdot s\cdot t\cdot u$$

As you can see, we have four equations with our four roots as variables, and that our four variables form a certain set of polynomials that are special, because they are called ***symmetric polynomials***. Symmetric polynomials share an interesting property -- if we exchange two variables in the above equations, the polynomials stay the same. You can try to check this property by exchanging two roots in the above equations, let us say ***r*** and ***s.*** You will notice that if you do this, the polynomials will not change.

However, these kinds of polynomials do not only have this property. Actually they have other important properties. For example, the above decomposition is usually called the normal form of the polynomial which can be done to any polynomial. A normal form will always be a set of symmetric polynomials. The above formulas are actually called ***Viete's formulas*** which were named after French mathematician ***François Viete***.

Symmetric polynomials don't seem to be a great deal. However, they are very important for ***Galois Theory***, which we will not study in this book since Galois Theory has to do more with abstract algebra than with linear algebra. Galois Theory is very important today because many applications use it to simplify situations.

History of Evariste Galois

Evariste Galois *was a French mathematician who lived during the first half of the nineteenth century. At a very young age Galois had mastered several topics in mathematics, but was rejected from an academy* *because of his political career, which sometimes even got him into jail. Galois, through his studies, was able to give an immense jump in theory of equations, solving problems that had been wondering mathematics for at least three centuries.*

Probably the most dramatic fact about Galois was that he died at a very young age, when he was 20 years old. He died in a duel with another man, the reason is unknown but it is believed that it was because of a woman and an honor debt. The particular characteristic about Galois death is that, the night before he died, afraid of his life, he wrote a very detailed paper with his discoveries and lent them to his friends asking them to publish it, which they finally did, 14 years after Galois death.

Going back to our topic, you might be very excited to try and solve for the roots of our polynomial based on its normal form. The truth is, if we try to do this, we will be going back to original polynomial. However, this is not a reason to be upset. These polynomials have other applications than Galois theory, like symmetry in multivariate polynomials which is a very desired condition.

Another Bound on the Roots of Polynomials

Symmetric polynomials are also useful in giving another bound on the roots of polynomials. Let us look one moment at the normal form expressed before. In case our roots are named $x_1, x_2, \ldots, x_n$, then we can see that if we have a general n^{th} degree polynomial of the following form:

$$p(x) = a_0 + a_1 \cdot x + a_2 \cdot x^2 + a_3 \cdot x^3 + \ldots + x^n$$

Then, through Viete's formulas:

$$a_1 = \sum_{i=1}^{n} x_i \qquad a_2 = \sum_{\substack{i=1 \\ j>i}}^{n} x_i \cdot x_j$$

Now let us look at some statistical study. In statistics, usually there is a set of data, like our roots, and what is normally done in statistics is to get the mean and variance of that set of data. Using our roots, let us show how the mean $\overline{x}$ and variance **s** are calculated:

$$\overline{x} = \frac{\sum_{i=1}^{n} x_i}{n} \qquad s^2 = \frac{1}{n}\sum_{i=1}^{n}(x_i - \overline{x})^2 = \frac{1}{n}\sum_{i=1}^{n}(x_i)^2 - \overline{x}^2$$

In statistics, it is shown by ***Samuelson's inequality*** that for any $\boldsymbol{x_i}$, the following must be accomplished:

$$\overline{x} - s\cdot\sqrt{n-1} \le x_i \le \overline{x} + s\cdot\sqrt{n-1}$$

However, we can correlate this statistical approach to the roots of our polynomial. For example, we can see that:

$$\overline{x} = \frac{a_1}{n} \qquad s^2 = \frac{a_1^2 - 2\cdot a_2}{n} - \left(\frac{a_1}{n}\right)^2$$

Therefore, we just found another bound for the roots of our polynomial through the above inequality. However, this approach is not only because of statistics. Edmond Laguerre also made this derivation during the nineteenth century, however, he did not notice the statistical characteristic of the problem. Nevertheless, the above inequality is also known as the ***Laguerre-Samuelson inequality.***

THE EXPONENTIAL MATRIX

So far, we have studied univariate polynomials and talked a little bit about symmetric multivariate polynomials. However, we did this variables that are scalar quantities in nature, but actually variables can also be matrices. We will show this situation in two cases, one using the ***exponential matrix*** and the other using the ***Cayley-Hamilton theorem.***

The exponential matrices are usually present in solving sets of systems of differential equations. An example of this is the following:

$$\vec{x}' = A\cdot\vec{x}$$

where vector $\boldsymbol{x}$ is the vector we are trying to find. You have probably encountered several problems involving differential equations. But for now, let us make the assumption that we have already solved the above equation with integration:

$$\frac{d\vec{x}'}{\vec{x}} = A\cdot dt$$

$$\vec{x} = \overrightarrow{x_0} \cdot e^{A \cdot t}$$

Where vector $\boldsymbol{x_0}$ are initial conditions. The above equation may seem weird because how are we able to take the $\boldsymbol{A}^{\text{th}}$ power of $\boldsymbol{e}$ when $\boldsymbol{A}$ is a matrix? The answer to this question resides in the Taylor series of the exponential matrix. Let us remember how that looks like:

$$e^x = 1 + x + \frac{x^2}{2!} + \frac{x^3}{3!} + \ldots + \frac{x^n}{n!}$$

Let us substitute $\boldsymbol{x}$ by matrix $\boldsymbol{A \cdot t}$ in the above expression. If we do this we get:

$$e^{A \cdot t} = I + A \cdot t + \frac{(A \cdot t)^2}{2!} + \frac{(A \cdot t)^3}{3!} + \ldots + \frac{(A \cdot t)^n}{n!}$$

The above expression demonstrate that we are able to take the $\boldsymbol{A}^{\text{th}}$ power of $\boldsymbol{e}$ where $\boldsymbol{A}$ is a matrix, since the result will also be another matrix. However, the above expression is not the most commonly used since it can be reduced even more. Assuming that matrix $\boldsymbol{A}$ has an eigendecomposition like the following:

$$A = V \cdot \Lambda \cdot V^{-1}$$

Then we can input this eigendecomposition into our last equation and get:

$$e^{A \cdot t} = I + V \cdot \Lambda \cdot V^{-1} \cdot t + \frac{(V \cdot \Lambda \cdot V^{-1} \cdot t)^2}{2!} + \frac{(V \cdot \Lambda \cdot V^{-1} \cdot t)^3}{3!} + \ldots + \frac{(V \cdot \Lambda \cdot V^{-1} \cdot t)^n}{n!}$$

Remember that $\boldsymbol{t}$ is just a constant. If we remember from the eigenvalue chapter we know that the above powers are reduced in the following way:

$$e^{A \cdot t} = I + V \cdot \Lambda \cdot t \cdot V^{-1} + \frac{V \cdot (\Lambda \cdot t)^2 \cdot V^{-1}}{2!} + \frac{V \cdot (\Lambda \cdot t)^3 \cdot V^{-1}}{3!} + \ldots + \frac{V \cdot (\Lambda \cdot t)^n \cdot V^{-1}}{n!}$$

And if we factor $\boldsymbol{V}$ and its inverse at the left and right side, respectively, we will get:

$$e^{A \cdot t} = V \left[I + \Lambda \cdot t + \frac{(\Lambda \cdot t)^2}{2!} + \frac{(\Lambda \cdot t)^3}{3!} + \ldots + \frac{(\Lambda \cdot t)^n}{n!} \right] V^{-1}$$

As you can see, the infinite sum in the middle of the above equation represents another Taylor expansion. Notice that

$$I + \Lambda \cdot t + \frac{(\Lambda \cdot t)^2}{2!} + \frac{(\Lambda \cdot t)^3}{3!} + \ldots + \frac{(\Lambda \cdot t)^n}{n!} = e^{\Lambda \cdot t}$$

However, this represents the same problem since we have a base elevated to a matrix, but in this case we have the incredible advantage that matrix Λ is diagonal. By knowing that matrix Λ is diagonal it is not difficult to show that if

$$\Lambda = \begin{pmatrix} \lambda_1 & 0 & \cdots & 0 \\ 0 & \lambda_2 & \cdots & 0 \\ \vdots & \vdots & \ddots & \vdots \\ 0 & 0 & \cdots & \lambda_n \end{pmatrix}$$

,then

$$e^{\Lambda \cdot t} = \begin{pmatrix} e^{\lambda_1 \cdot t} & 0 & \cdots & 0 \\ 0 & e^{\lambda_2 \cdot t} & \cdots & 0 \\ \vdots & \vdots & \ddots & \vdots \\ 0 & 0 & \cdots & e^{\lambda_n \cdot t} \end{pmatrix}$$

Completing the procedure. However, all this is valid in case matrix ***A*** has an eigen-decomposition, but what happens if it does not? In that case what is usually done is to take several powers of ***A*** and truncate the infinite sum when the error passes below certain level, or, in case ***A*** is a Nilpotent matrix (a Nilpotent matrix is a matrix that has all zero entries after taking a certain number of its powers), then truncate the sum when A^n becomes the zero matrix. Another approach in case ***A*** does not have an eigen-decomposition, somewhat more difficult, is to calculate the Jordan Normal form of ***A*** and do the same approach as we did before.

Cayley-Hamilton Theorem

The Cayley-Hamilton theorem is a very important theorem not just in Linear Algebra, but also in abstract algebra and topology. The Cayley-Hamilton theorem was developed by ***Arthur Cayley***, Britain's leading pure mathematician of the 19th century, with great contributions to abstract algebra like the Cayley tables (also by ***Rowan Hamilton.*** The same Rowan Hamilton that discovered quaternions and had many contributions to sciences like physics, astronomy and mathematics).

This theorem states that every square matrix satisfies its own characteristic polynomial. From the last topic we saw that we are able to make polynomials not only with scalars but also with matrices, therefore, let us look at what the theorem is telling us. Suppose we have a square matrix ***A*** and that its characteristic polynomial is the following:

$$det(A-\lambda I)=\sum_{i=0}^{n} c_n \cdot \lambda^n$$

Where the different c_i constants are contained within a field of scalars. If we input our parameter lambda as ***A*** in the above equation, we would get that:

$$det(A-A)=\sum_{i=0}^{n} c_n \cdot A^n = det(0)=0$$

wherein we can see that matrix ***A*** does satisfy its own characteristic polynomial. This fact may not seem surprising but it has many applications and uses that are just too many to enumerate. For now we are interested in the fact that in a way the above equation allows us to get the inverse of ***A***.

Let us look at this in the following way. Suppose, that by some method, we are able to get the characteristic polynomial of our matrix ***A*** and we input ***A*** as our lambda. This would look like the following expression:

$$c_0 + c_1 \cdot A + c_2 \cdot A^2 + \ldots + (-1)^n \cdot A^n = 0$$

Remember that c_0 is the determinant of ***A***, however, if we multiply the above expression by the inverse of ***A***, and solve for A^{-1}, we would get

$$A^{-1} = \frac{c_1 + c_2 \cdot A + \ldots + (-1)^n \cdot A^{n-1}}{-c_0}$$

Take note that we have found a way to correlate the inverse of ***A*** with the powers of ***A***. This is one of usefulness of the Cayley-Hamilton theorem. It also allows us to express the powers of a matrix in terms of that matrix. It is true that we saw how to take the nth power of a matrix through its eigendecomposition, but in cases when ***A*** does not have an eigendecomposition we can use the Cayley-Hamilton theorem. Let us look at a simple example to show these two out of the many properties of this theorem. Suppose we have the following matrix ***A***:

$$A = \begin{pmatrix} 1 & 2 \\ 3 & 4 \end{pmatrix}$$

At a first sight we can notice that its characteristic polynomial is: $det(A - \lambda \cdot I) = \lambda^2 - 5\lambda - 2$

If we input ***A*** in the above expression we would get that:

$$A^2 - 5 \cdot A - 2 \cdot I = 0$$

From here we can calculate the inverse of ***A*** like this:

$$A^{-1} = \frac{A - 5 \cdot I}{2}$$

Or we can calculate the powers of matrix ***A*** like this:

$$A^2 = 5 \cdot A + 2 \cdot I$$

$$A^3 = 5 \cdot A^2 + 2 \cdot A = 5(5 \cdot A + 2 \cdot I) + 2 \cdot A = 27 \cdot A + 10 \cdot I$$

$$A^4 = 27 \cdot A^2 + 10 \cdot A = 27(5 \cdot A + 2 \cdot I) + 10 \cdot A = 145 \cdot A + 54 \cdot I$$

Which is an amazing way to solve for the powers of our matrix ***A*** in case it does not have an eigendecomposition (notice the geometrical progression). However, the question arises, how do we compute the characteristic polynomial of a matrix without calculating its determinant? Luckily, iterative methods exist for this situation, and the most common of them is the Faddeev-Leverrier method, named after Dmitry Konstantinovich Faddeev and Urbain Leverrier. The latter is a famous mathematician who discovered planet Neptune through mathematics.

CHAPTER 9

AREAS CONCERNING LINEAR ALGEBRA

"The reward of a thing well done is having done it"

Ralph Waldo Emerson

INTRODUCTION

Reward, a very nice word. This is the final chapter of this book, the final chapter of a long journey that started some hundreds of pages ago. As the author, I would like to thank you for reading up to this point, because that means you like this piece of work and it served as an aid to increase your knowledge. For this reason, this chapter is especially dedicated to enjoy and appreciate many facts about linear algebra that usually get omitted in more professional books, or are just whispered briefly during the final moments of a math class.

These facts are important, because, as we shall see, they form part of history and, in some occasions, give other tools to solve problems in a sort of amusing way. Besides, one quote at the beginning of this book said that scientists study nature because it is beautiful, and nature is beautiful because we delight in it. While studying and maturing our algebraic consciousness, we take into account that algebra is an incredible body of knowledge that is beautiful in itself, and is a powerful being since it can be used to tackle several problems. For this reason also, in this chapter, we shall discuss and see situations, problems, and techniques that are very ingenious ways to model certain requirements. However, this chapter will not contain many exercises, problems, or complicated equations. It is dedicated to putting into use the many tools that we studied along the way in situations where these tools will let us go further in our search for knowledge. Let us begin.

BASEL PROBLEM

The Basel problem was a mathematical problem that arose in 1644 when posed by Pietro Mengoli, an Italian mathematician and clergyman. The Basel problem is a very simple problem but not so simple to solve since it was opposing mathematicians until 1734 when ***Leonhard Euler***, a very famous mathematician who made several contributions to mathematics, algebra and physics, managed to solve it. The Basel problem asks for the precise summation of the reciprocals of the squares of the natural number. In other words, it is an infinite series that asks for the following result:

$$\sum_{n=1}^{\infty}\frac{1}{n^2}=\frac{1}{1^2}+\frac{1}{2^2}+\frac{1}{3^2}+\ldots$$

This sum is demonstrated to converge to 1.644934…, however, we still need to prove this number. There are three common methods to solve the Basel problem. One of the methods even uses Fourier Series, but for our proof, we will use the method which Euler used, which is the most direct method. For this proof we need to take something into account. Euler was very familiar with the Taylor series of trigonometric functions. He also got his famous Euler's Identity $e^{i\pi}=-1$ using the Taylor Series.

Now let us carry on with the proof. If we remember from a basic Calculus course, we know that the Taylor Series of the sine function is:

$$\sin(x)=x-\frac{x^3}{3!}+\frac{x^5}{5!}-\frac{x^7}{7!}\ldots$$

If we divide the above polynomial by $\boldsymbol{x}$, we would get

$$\frac{\sin(x)}{x} = 1 - \frac{x^2}{3!} + \frac{x^4}{5!} - \frac{x^6}{7!} \ldots$$

From the fundamental theorem of algebra we know that the above function has the same zeros as the zeros of $\sin(x)$. This allows us to factor the above polynomial in terms of its roots. The reader should know that a zero of $\sin(x)$ is any integer multiple of π. Therefore,

$$\frac{\sin(x)}{x} = \left(1 - \frac{x}{\pi}\right) \cdot \left(1 + \frac{x}{\pi}\right) \cdot \left(1 - \frac{x}{2\pi}\right) \cdot \left(1 + \frac{x}{2\pi}\right) \cdot \left(1 - \frac{x}{3\pi}\right) \cdot \left(1 + \frac{x}{3\pi}\right) \ldots$$

$$\frac{\sin(x)}{x} = \left(1 - \frac{x^2}{\pi^2}\right) \cdot \left(1 - \frac{x^2}{4\pi^2}\right) \cdot \left(1 - \frac{x^2}{9\pi^2}\right) \ldots$$

If you remember the theory we saw on symmetric polynomials, you may have noticed that we have the above polynomial in terms of its roots. Therefore, if we expand the above polynomial we should get the same polynomial as when we started. If we hypothetically do this and take all terms with x^2, then we must equal the result of all of those terms and equate it to the term corresponding with x^2 in the original polynomial. If we do this we would get that:

$$\frac{-1}{3!} = -\left(\frac{1}{\pi^2} + \frac{1}{4\pi^2} + \frac{1}{9\pi^2} + \ldots\right) = -\frac{1}{\pi^2} \cdot \sum_{n=1}^{\infty} \frac{1}{n^2}$$

From here we can get the solution to the Basel problem by solving for the summation on the right. If we do this we would finally get

$$\sum_{n=1}^{\infty} \frac{1}{n^2} = \frac{\pi^2}{6} \approx 1.644934\ldots$$

which completes the proof and solution of the Basel problem. If you are is wondering where the Basel problem got its name, it came from Euler's hometown Basel, which still exists today in Switzerland. This problem, even when it was presented as a single problem, gave rise to other problems that can be generalized into a single question, that is, *"is it possible to compute other summations of the same style as the Basel problem but with different exponents?"* In other words, if it is possible to calculate summations of the following form, then

$$x = \sum_{n=1}^{\infty} \frac{1}{n^s} \rightarrow s = 1, 2, 3 \ldots$$

and the answer is "yes" in most cases. It is possible to calculate the above summation for every ***s*** greater than 1, and this problem was presented by Euler himself. Euler then also proved that when ***s*** is an even number the above summation gives results of the following form:

$$\sum_{n=1}^{\infty} \frac{1}{n^s} = \frac{\pi^s}{c} \rightarrow when\, s\, is\, even$$

where ***c*** is a constant that varies according to the exponent. For example, in case ***s*** is 4 we would get

$$\sum_{n=1}^{\infty} \frac{1}{n^4} = \frac{\pi^4}{90}$$

RIEMANN ZETA FUNCTION

Bernhard Riemann, a German mathematician during the 19th century, was the father of the Riemann Zeta function. The Riemann Zeta Function is the same function as the one that Euler presented, however, Riemann expanded it with complex analysis giving the possibility that ***s*** is a complex number. The proper name of this function is the Riemann Zeta function, however, sometimes it is also called as the Euler-Riemann Zeta function, which is usually presented as

$$\zeta(s) = \sum_{n=1}^{\infty} \frac{1}{n^s}$$

where ***s*** can be any complex number. It is of no surprise that the Riemann Zeta function converges for any number ***s*** with real part higher than 1. However, this did not stop mathematicians in doing an analytical continuation on the complex plane to include any possibility of s**.** In case you are wondering, let us check what happens if we input a value of ***s*** equal to -1.

$$\zeta(-1) = \sum_{n=1}^{\infty} \frac{1}{n^{-1}} = 1 + 2 + 3 + 4 + \ldots$$

Looks familiar? It should be! We saw this summation at the beginning of chapter 2, and we said that the above summation has a value equal to $-1/12$, however, this is not the only interesting thing about this function. The Riemann Zeta function has many particular points along the whole complex plane. For example, it has infinite trivial zeros at every negative even integer, it is shown to be only divergent at a single point (which corresponds to $s=1$, yielding the Harmonic series which is shown to be divergent), and there is a hypothesis with this function, known as a the Riemann Hypothesis, which is one of the Millennium problems and also one of the Hilbert's eighth problem.

You may be wondering what is so especial about the Riemann Zeta function, and you should be. This function has special applications in number theory, especially relating to the prime numbers, or the exact number of primes in a certain range. This function also would be very useful to prove several conjectures in mathematics, like Goldbach's Conjecture which is very important in number theory.

The field of study concerning this infinite sums is very wide, and the deeper we get into it the more interesting it gets. For example, from these approaches with the Riemann Zeta function we are able to get constants that appear in many places in physics like the Apery's constant and the Euler-Mascheroni constant; and we are able to get useful relations with particular numbers like π or e.

Speaking about π or e, now would be a good time to talk about a particular set of numbers in which these two are involved. These are the transcendental numbers.

TRANSCENDENTAL NUMBERS

Transcendental numbers are a special set of numbers that belong in mathematics, and we will see why special attention is given to these set of numbers. But first, let us look at what a transcendental number is. A transcendental number is a number that is not algebraic, and an algebraic number is a number that is a root to a certain polynomial with integer (or rational) coefficients. Let us analyze this last expression.

We said that any transcendental number is not an algebraic one, so we need to understand what an algebraic number is in order to get to what we want. We said that an algebraic number must be a root to a certain polynomial with integer or rational coefficients, so let us look at some examples to understand this expression. 7 is an algebraic number, because 7 is a root of the following polynomial:

$$p(x) = x - 7$$

Where we see we have a polynomial with integer coefficients. The reader, from the above example, can deduce that any integer number is also an algebraic number because any integer number will also be a root to a certain polynomial with integer coefficients. Rational numbers are

also algebraic. To see this, assume that we want to prove that $\frac{1}{2}$ is an algebraic number. One-half is a root of the following polynomial:

$$p(x) = 2x - 1$$

As you can see we have a polynomial with integer coefficients. From the above expression the reader can also deduce that any rational number is also an algebraic number. Now let us look at the irrational numbers and let us try to prove which irrational numbers are also transcendental. One of the most famous irrational numbers is $\sqrt{2}$, but $\sqrt{2}$ is also algebraic, because it is a root of the following polynomial:

$$p(x) = x^2 - 2$$

Another famous irrational number that is also algebraic is the golden ratio (which is actually the most irrational number). In case the reader does not remember, the golden ratio is given by the following number, and it is a root of the following polynomial:

$$\phi = \frac{1+\sqrt{5}}{2} \rightarrow p(x) = x^2 - x - 1$$

However, not all irrational numbers are algebraic, why? Because irrational numbers have an infinite number of decimal places that do not follow any pattern. However, in some cases, that chaotic succession of decimal places can be expressed by a root of a polynomial, but in other cases they cannot, and from here is where transcendental numbers got their name.

Transcendental numbers are called that way because they are a set of numbers that cannot be expressed under any finite methodology we could imagine. They are a set of numbers that transcend mathematics, and we can only express the idea of this kind of numbers by already having that idea within mathematics. This is somehow difficult to understand, but let us check another difference between algebraic and transcendental numbers to understand this.

Let us start with the basic set of numbers, ***the integers***. To build the entire set of integers we only need number 1 and the addition and subtraction operation, since any integer number is the sum or difference of two integers. For example, $2 = 1 + 1$, therefore, we obtained a new element which is 2. From here we can build another integer like this: $3 = 2 + 1$, obtaining a new element, and so on with all integers, just using addition and subtraction.

Let us step up to ***the rational*** numbers. If we now include the operations of multiplication and division, we are able, with help of the integers, to get any rational number. Remember that since we managed to get any integer number with addition and subtraction, then with multiplication and division we will be able to get any rational number, since any rational number is the quotient of two integers.

Stepping up again, we now get to ***the irrational*** numbers. It is impossible to get an irrational number only with the four basic arithmetic operations of addition, subtraction, multiplication and division. Irrational numbers come by including the operations of taking powers and roots of the rational numbers with the help of the four basic arithmetic operations. For example, $\sqrt{2}$ is the square root of a rational number (which is also an integer).

Now let us look at something very interesting. We know that any transcendental number must also be irrational. However, not all irrational numbers are transcendental, so let us divide the irrational numbers into two: ***algebraic*** and ***transcendental***. The set of algebraic numbers is limited since it only uses the rational numbers as bases and takes rational roots of those rational numbers. That is, any irrational algebraic number has the following form:

$$x = \sqrt[b_1]{a_1} + \sqrt[b_2]{a_2} + \ldots + \sqrt[b_n]{a_n} \rightarrow where a_1, a_2, \ldots, a_n, b_1, b_2, \ldots, b_n are rational$$

where ***n*** is a finite number. This last expression gives space for the transcendental numbers to appear, because there exist numbers that cannot be expressed under the above description due to the "restriction" of using only rational numbers as bases and exponents. Now that we have a very good understanding of what makes a transcendental number transcendental, it would very nice to talk about their importance and how they got into mathematics.

We said that we cannot express the idea of a transcendental number without the transcendental number itself, and it is true. Let us look at this by taking π as an example. We are not able to express or even write π using integers, rational numberss, or as the nth root of another number, so the only way we have to express the idea of π is by already having the idea of π. The same is true for any transcendental number.

There is a lot of philosophical discussion from the last idea -- whether math was created or discovered, however, we are not here to talk about that. What we are going to talk about here now is how to compute transcendental numbers and their importance. To notice this importance, we must remember, from what we have mentioned before, that transcendental numbers cannot be computed with a finite approach. However, they can be computed, in some occasions, with an infinite approach, like π and ***e*** in the following expressions:

$$\pi = \sqrt{6 \cdot \sum_{n=1}^{\infty} \frac{1}{n^2}} \qquad e = \lim_{x \to \infty} \left(1 + \frac{1}{x}\right)^x$$

This has great consequences in the computation of transcendental number, because neither computers nor humans can work with an infinite number of decimals, or an infinite number of steps to calculate a number, especially when such special numbers appear almost anywhere in mathematics. For example, π usually appears where there is a circle, in the Pascal's triangle, engineering, physics, complex analysis, geometry, and more. The same is true with ***e***, which has great applications in economy, too.

This impossibility of computation gives rise to approximation errors, and makes mathematicians, and us, to work with errors as a common part of our lives. That is the importance of transcendental numbers. They are very, immensely I should say, important numbers that appear almost in any field of study, but we must always deal with the error behind their computation and significant figures. You might be asking why they so important. For that reason, we will proceed with another topic on this chapter which is error analysis.

ERROR ANALYSIS

During the course of this book, we have focused mainly about theorems, applications, proofs and examples of problems and situations that may arise during our daily lives as mathematicians. However, it was omitted on purpose the part on working with errors. It was silently omitted because in almost all calculations we used four significant figures, but that is not an excuse not to pay attention to errors.

Error analysis focuses on the study of dealing with how good or how accurate our solution is. The reasons behind this very well based and great field of study are that, in some occasions, errors tend to cause so much trouble even when they are minimum, and that errors are with us all day -- in every measurement, every time we don't take complete decimals, every time we use a transcendental number and more.

In this topic, we will only discuss about errors in computation, meaning errors that exist in the abstract world of number theory and mathematics. We will not discuss errors that exist in the real world since they are something else that physicist and statisticians usually deal with. Having said this, let us discuss how errors affect our calculations and our results.

When we compute numbers, coefficients, and polynomials, we usually we take approximations of values for several reasons like easement and memory usage. However, in some cases we must be careful with the approximations we make. Let us look at a simple example to show this. Suppose we have a number and we must take the square of that number, something very straightforward like this:

$$y(x)=x^2$$

Let us assume that our real value of $\boldsymbol{x}$ is 2, but we used an approximate value of 2.01. If we square both numbers and get the error, we would get:

$$y(2)=4; y(2.01)=4.0401 \rightarrow e=y(2)-y(2.01)=-0.0401$$

0.01 don't seems much of an error right? Neither does 0.0401, however, what happens if our real value of x is 1000 and our approximation is 1001? Then,

$$y(1000)=1,000,000; y(1001)=1,002,001 \rightarrow e=y(1000)-y(1001)=-2001$$

Although 1 does not seem much in comparison to 1000, the error now is 2001. That may seem little compared to a million. However, imagine if that million represents a debt, or heart diseases patients, or any other important magnitude that may come in mind. In that case, then we should be very critical with our error. However, that is not the only point we want to show with the last example; we also want to show that in some occasions our error will be affected by the operations we are making and by the magnitude of the values we are taking. Comparing both cases, with one case squaring 2 and the other case squaring 1000, they both square a number, but the second case gives more error than the first one, which is because 1000 is much bigger than 2.

However, squaring a number is not the only possible operation. Addition, subtraction, multiplication, division, integration and differentiation are also possible. The reason we are explaining this is because computation is necessary in our daily life situations, and if the reader desires to be a well based algebraist then he must know how operations and magnitudes affect the error in his results. To see this, suppose we have a certain value $\boldsymbol{a}$ and another value $\boldsymbol{b}$, with their respective errors $\Delta a$and Δb, assuming that both errors are always positive. Now let us assume we make all possible operations between both numbers. Therefore,

- Addition and subtraction:

$$(a+\Delta a)+(b+\Delta b)=(a+b)+(\Delta a+\Delta b) \rightarrow error=\Delta a+\Delta b$$

$$(a+\Delta a)+(b+\Delta b)=(a+b)+(\Delta a+\Delta b) \rightarrow error=\Delta a+\Delta b$$

While adding or subtracting numbers we can see that the resulting error will only depend on the error of the initial terms we had. By the way, in case the reader is wondering in the subtraction process, no, there is no mistake, the error is $\Delta a + \Delta b$ when subtracting ***a*** and ***b***. The reason why we add the errors and not subtract them is because, in reality we do not have the exact quantity of the error, therefore, we need to give an upper bound for it. In case the errors would be subtracting like $\Delta a - \Delta b$, then that would mean that the error is decreasing. However, we cannot assure that, therefore, we treat the operation so that the bound of our error always increases.

- Multiplication

$$(a+\Delta a)\cdot(b+\Delta b)=(a\cdot b)+(b\cdot\Delta a)+(a\cdot\Delta b)+(\Delta a\cdot\Delta b)\rightarrow error=b\cdot\Delta a+a\cdot\Delta b+\Delta a\cdot\Delta b$$

While multiplying our values ***a*** and ***b***, we noticed that our error is proportional to the values of ***a*** and ***b*** and to the errors of ***a*** and ***b.*** From the above equation we can notice that, in case the value of ***a*** is big then the error of ***b*** must be small for the total error to be small, and vice versa with the value of ***b*** and the error of ***a***. We can also notice that we need to be careful in case the values of ***a*** and ***b*** are very small. In that case, their errors must be even smaller for the total error to be acceptable.

- Division

$$\frac{a+\Delta a}{b+\Delta b}=\frac{a}{b+\Delta b}+\frac{\Delta a}{b+\Delta b}\rightarrow error\approx\frac{\Delta a}{b+\Delta b}$$

In case of division, we must also be very accurate with errors. We see that the total error is proportional to the error of ***a*** and inversely proportional to the magnitude of ***b*** and its error. Meaning that if we want to minimize the total error then the error of ***a*** must be very small, and in case the magnitude of ***b*** is small then the error of ***b*** must also be very small.

- Exponentiation

$$(a+\Delta a)^n=a^n+(n-1)\cdot a^{n-1}\cdot\Delta a+\ldots(\Delta a)^n\rightarrow error>(n-1)\cdot a^{n-1}\cdot\Delta a+(\Delta a)^n$$

In case of exponentiation, we see that the error of taking the nth power of ***a*** is proportional to the (***n***-1) power of ***a***, therefore, we can see that in case the magnitude of ***a*** is big, then the value of the error of ***a*** must be very small to get an acceptable value of the total error. Also in case the magnitude of ***a*** is very small, then the error of ***a*** must also be very small so that the total error is acceptable.

- Integration

$$\int a + \Delta a = \int a + \int \Delta a \rightarrow error = \int \Delta a$$

In this case let us assume that ***a*** is a function. Concerning integration, we see that the error is equal to the integral of the error of ***a***. In case the interval of integration is big, then the error of ***a*** must be very small to get an acceptable result. We can see this in this image:

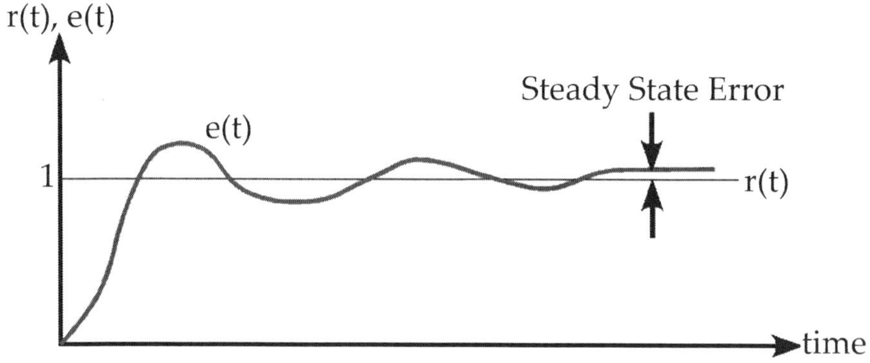

Figure 27: Steady state error.

Source: dcontrolsystems.com

-Differentiation

$$\frac{d}{dx}(a + \Delta a) = \frac{da}{dx} + \frac{d(\Delta a)}{dx} \rightarrow error = \frac{d(\Delta a)}{dx}$$

In case of differentiation, then our error must not necessarily be small, rather than constant. The error must not vary greatly because the derivative of the error can be greatly big. The ideal case while deriving a function is that the error of that function is constant so that $\frac{d(\Delta a)}{dx} = 0$. This can be seen in the following image:

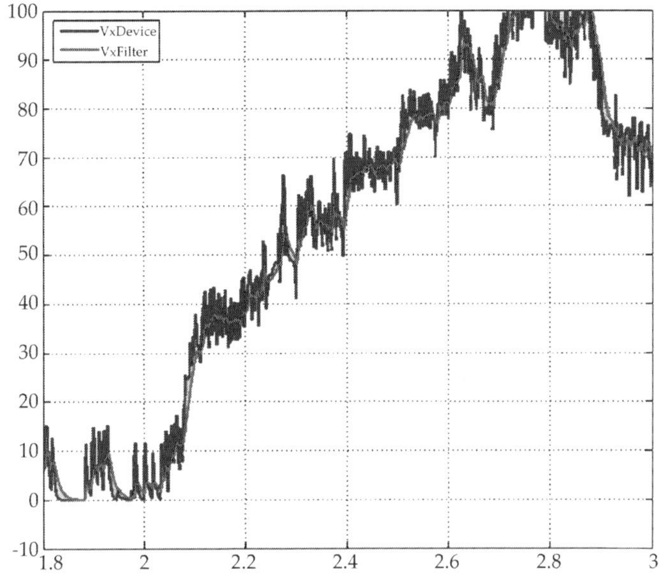

Figure 28: Derivative error.

Source: i.stack.imgur.com

This is a common image of a signal processing situation. If we try deriving the blue function shown above, the error will be enormously great because of the great variance, however, if we try deriving the red function the error shall be smaller because it has less variance and its derivative will be more stable.

This is how the operations we make alter the error through our calculations. However, this is not the only way to consider the error during our calculations. There are other parameters that we have mentioned throughout the course of this book, especially with matrices when we talked about the condition number of a matrix. If you will recall, the condition number of a matrix is a measure of how much our system is ill-conditioned, that is, how much our results will change by small changes in the coefficients of our matrix.

It would be logical to assume that small changes in the coefficients of a matrix should give small changes in the solution to our system, but that is not always true since this will depend on the condition number of our matrix. Let us look at the following systems and its disturbed versions, and look at the difference in the solutions:

$$\begin{pmatrix} 10 & 7 & 8 & 7 \\ 7 & 5 & 6 & 5 \\ 8 & 6 & 10 & 9 \\ 7 & 5 & 9 & 10 \end{pmatrix}\begin{pmatrix} x_1 \\ x_2 \\ x_3 \\ x_4 \end{pmatrix} = \begin{pmatrix} 32 \\ 23 \\ 33 \\ 31 \end{pmatrix}$$

$$\begin{pmatrix} 10 & 7 & 8.1 & 7.2 \\ 7.08 & 5.04 & 6 & 5 \\ 8 & 5.98 & 9.89 & 9 \\ 6.99 & 4.99 & 9 & 9.98 \end{pmatrix}\begin{pmatrix} x'_1 \\ x'_2 \\ x'_3 \\ x'_4 \end{pmatrix} = \begin{pmatrix} 32 \\ 23 \\ 33 \\ 31 \end{pmatrix}$$

$$\begin{pmatrix} 10 & 7 & 8 & 7 \\ 7 & 5 & 6 & 5 \\ 8 & 6 & 10 & 9 \\ 7 & 5 & 9 & 10 \end{pmatrix}\begin{pmatrix} x''_1 \\ x''_2 \\ x''_3 \\ x''_4 \end{pmatrix} = \begin{pmatrix} 32.1 \\ 22.9 \\ 33.1 \\ 30.9 \end{pmatrix}$$

We would expect that such little variation of the coefficient matrix and the independent term does not change much the result, but indeed they do change like this:

$$\begin{pmatrix} x_1 \\ x_2 \\ x_3 \\ x_4 \end{pmatrix} = \begin{pmatrix} 1 \\ 1 \\ 1 \\ 1 \end{pmatrix} \qquad \begin{pmatrix} x'_1 \\ x'_2 \\ x'_3 \\ x'_4 \end{pmatrix} = \begin{pmatrix} -81 \\ 137 \\ -34 \\ 22 \end{pmatrix} \qquad \begin{pmatrix} x''_1 \\ x''_2 \\ x''_3 \\ x''_4 \end{pmatrix} = \begin{pmatrix} 9.2 \\ -12.6 \\ 4.5 \\ -1.1 \end{pmatrix}$$

Notice how this phenomenon happens even if we change the independent term of our system of equations. However, it must be noted that this dramatic change in our results happen because the condition number of our matrix is dramatically big, or in other terms, our matrix is very close to being non-invertible.

This is the case for matrices, however, there is also ill-conditioning for matrices. This was, again, omitted in last chapter, but this is a very important topic in calculating the roots of a polynomial. The most relevant example that is usually given by several authors in this topic is Wilkinson's polynomial, which was presented in 1963 by James Wilkinson, an English numerical analyst. ***Wilkinson's polynomial*** is a very good example of ill-conditioned polynomials. Let us look at it with higher detail. Wilkinson's polynomial is given by:

$$w(x)=(x-1)\cdot(x-2)\cdot\ldots\cdot(x-20)$$

wherein the roots are the integers from 1 to 20, which are very well separated roots. When we expand the polynomial we get the following:

$$w(x)=x^{20}-210x^{19}+20615x^{18}-\ldots+20!$$

The other terms of the polynomial are omitted for their immense values, however, what we are interested in is just the second term of the above polynomial, the one containing x^{19}. This case of ill-conditioning is described, personally, by Wilkinson as the most traumatic experience in his career as a numerical analyst, because if we decrease the second coefficient (the -210) by only a minimal amount like 2^{-23}, making it -210.0000001192, then this new polynomial changes from having 20 real roots to having 10 real roots and 5 conjugate pairs, somewhat distorted, of complex roots. This means that only a change up to the seventh decimal place of that coefficient changes the roots completely. To check whether a polynomial is ill conditioned at a certain range, then we must calculate the derivative of that polynomial the closest as possible to the root -- if the derivative of the polynomial is high, then that root is very sensitive to small changes in the coefficients, and if the derivative is small then that root is not very sensitive.

There is a lot of discussion over this topic, we will not get into much detail of it since it does not only require time but also experience. You should probably encounter problems like this to really know how much a single approximation can change the results greatly from what you would expect. However, for the remaining part of this chapter we will focus on another topics that can also be very useful to know, like the Shoelace formula.

THE SHOELACE FORMULA

The Shoelace formula is a mathematical algorithm to check for the area of an enclosed region formed completely by straight lines with known vertices in a plane. The Shoelace formula was described by Meister and Gauss during the eighteenth century, reason why it is sometimes called the Gauss formula. This formula, ironically speaking, got its name because its calculation resembles a shoelace, and to demonstrate how the shoelace formula works we will make a simple example. However, first we must take into account the only two conditions for the Shoelace formula to work.

- the region whose area we are calculating must be closed and formed by a straight line; and
- the borders that surround the region must not intersect themselves at any point.

Now that we have said this, we will demonstrate how to use the Shoelace formula with a simple example and then explain the idea behind it. Let us suppose we have the following closed region formed by the three points A (2,3), B (-3,1) and C (-1.5,-2.5), as shown in the figure below:

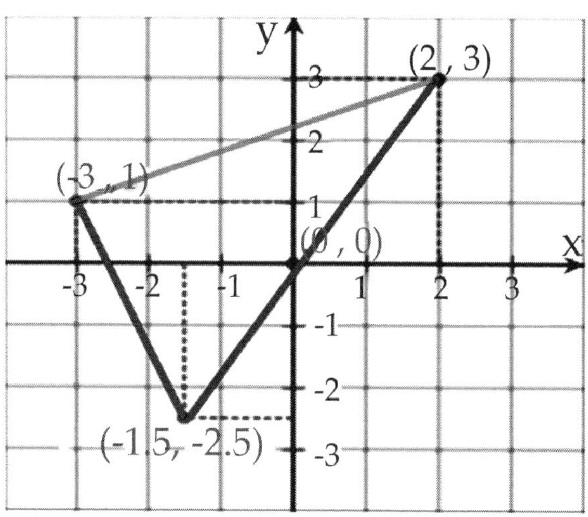

Figure 29: Example triangle.

Source: upload.wikimedia.org

We can solve the above problem in several ways -- we can do integration, apply geometry, going to theory on triangles, and so on. However, the shoelace formula is probably the simplest method to solve for the above situation. To apply the shoelace formula all we need to do is to list all our vertices with their components and cross-multiply consecutive coordinates of our vertices, then sum all those cross-multiplications and divide it by 2. Let us do this with the above triangle:

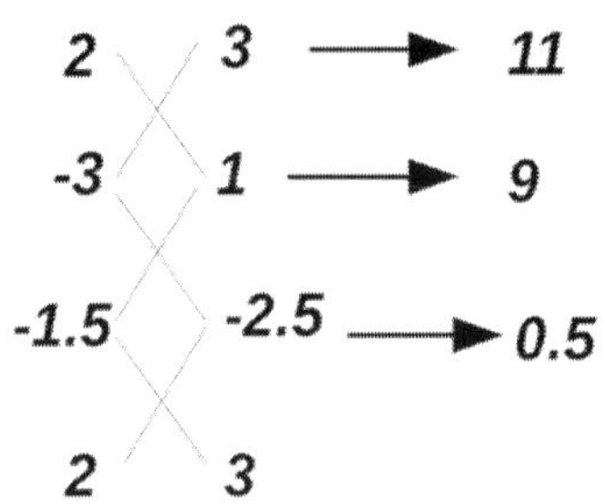

wherein we listed our 3 vertices and the first one at the end also. The result of the cross-multiplications are shown on the right of each component. In case you do not know how to do the cross multiplications, the first one would be $2 \cdot 1 - (-3) \cdot 3 = 11$ and so on. According to the formula we now must sum all cross-multiplications and divide by 2. Doing so would give us the enclosed area which is 10.25, which is the exact area.

Probably you are now wondering how this formula actually works. It works by calculating the area of triangles formed by consecutive points and the origin, and it does this by taking the cross product of the vectors that conform these triangles.

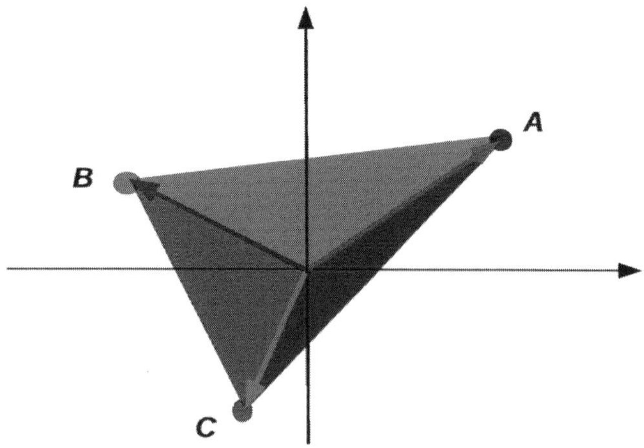

Figure 27: Shoelace formula background.

You can see that our former triangle is now formed by three triangles. These three triangles are formed by vectors whose components correspond exactly to the components of our vertices, therefore, each cross-multiplication corresponds to making a cross product between two vectors, meaning the area between those vectors. Then we sum all the areas and divide by 2 to get the area of the triangles (remember that a cross product calculates the area of the parallelogram). Relating to our former procedure we can see that the first cross-multiplication calculates twice the area of the green triangle, the second calculates twice the area of the red triangle, and the last twice of the area of the blue triangle.

This methodology can be seen easily because the origin is inside the area we are studying. The Shoelace formula would still work perfectly even if the origin is not within the region we are calculating (remember that a cross product can yield negative numbers, not because the area is negative but because of the vector orientations).

The Shoelace formula may seem very dull, but it has applications in several sciences and is of very good use in problems where lines are used to build objects, such as in game development or in 2D engineering drawings. If the reader wants to have a good knowledge in algebra and be well prepared to face problems in his career, he will need as many tools as he can get to tackle this kind of situations. The Shoelace formula can be also used in its 3D version to calculate volumes while building figures using planes (also very useful in 3D game development).

Believe it or not, game development is one of the greatest applications linear algebra has, however, it did not start there. Algebra began to enter the computational world, alongside numerical methods, by the optimization of networks, clusters and the storage and movement of information through distinct points of a net. Algebra is everywhere in a computer since every computational algorithm nowadays is highly probable to be optimized, approximated and implemented in a computer through an algebraic approach. Since this is the last chapter of the book it would be a nice moment to give last words as where the reader should go in case he wants to continue his formation as an algebraist.

In your continued pursuit of knowledge, this book recommends that you go for a course or read another book in *topology* in case you are interested in a wider generalization of geometry and space. In case you are interested on continuing with how information is processed and packed, then we recommend that you go for a course on *graph theory*, a truly incredible field related to algebra. And in case you are interested in sets, groups, fields, number theory and more, then we recommend that you go for higher level *abstract algebra* books and courses.

INDEX